ILLUSTRATED
BY
TWILLA JOHNSON

✹ ✦ ✹

PLANETS TO PULSARS:
A CITIZEN'S GUIDE TO
THE UNIVERSE

by
Jeff Dershem

Published by Dreamtime Press

3290 Valley View Drive
Grand Junction, Colorado 81503
970-462-7145
Email: Publisher@DreamtimePress.com

ISBN: 0-9823999-4-4

Dershem, Jeff

Planets to Pulsars, A Citizen's Guide to the Universe – 1st ed.

1. Textbook

DEDICATION

In memory of my friend and favorite physics professor, Dr. Gordon Gilbert. Thank you for providing such a shining example of a scientist and a human being. The universe is a little darker without you.

To Marvin and Dorothy, my parents, for putting up with me all of these years and for allowing me the freedom to just be me.

✴ ◆ ✴

TABLE OF CONTENTS

✸ ◆ ✸

INTRODUCTION

I know what you're thinking. Just what the world needs, one more book on astronomy (or any science for that matter) written for a general audience (i.e. the interested, yet mathematically or scientifically challenged reader). What's different about this one? What possessed the author to write it?

I'll answer the last question first. For several years I taught astronomy classes through the Community Education department of Mesa State College. One class was for beginning observers. I covered what features to look for when buying a pair of astronomical binoculars or a telescope, locating the constellations as a guide to finding all sorts of interesting objects in the night sky, and other topics. The second class, also for novices, was more of a brief history of the science of astronomy from the geocentric cosmology of Aristotle and Ptolemy to the discovery of and recent ideas on the nature of Dark Matter and Dark Energy.

My idea was to give the students a sense of what astronomers know and how they have come to know it. I also had the idea that it would be nice for the students to have a text that would complement my lectures, and fill in the gaps (there's only so much I can fit into a 1½ hour talk). Try as I might though, I couldn't find a book that fit the bill to my satisfaction. There seemed to be only one solution: write my own. So here we are.

What's wrong with all those other books? For one, most are filled with way too much trivial, "Ooh, Ahh," type information that contributes nothing to the reader's understanding of how science is done. Tables of the average distances of the various planets from the Sun, their average surface temperatures, and other such information can be found in most encyclopedias or on the internet. I would much rather leave the reader with an understanding of how astronomers determine the distances to the stars and the work that goes into such a project than present them with a list of the distances to the ten nearest stars. Science writers who try to impress their readers with pages of such tables (whether to add bulk because the publisher required a certain amount of pages, or for some other reason) sell them short.

Far too many basic science books contain endless biographical material on the scientists involved in a particular discovery. Not only does this break the flow of the narrative, it is not the place for such information in my opinion. The lives of scientists are at least as interesting as the lives of politicians, and sometimes certain aspects of a scientist's past can influence how he/she approached a problem, but copious biographical details (where the scientist was born, weight at birth, what flavor ice cream was his/her favorite, etc.) should be left to biographical works (there are many fine ones).

Quite a few basic science books try to cover it all. The basic process by which certain nebulae shine is the result of quantum physics, but entire books have been written about that subject alone. I do not propose to reproduce such works here. For those interested who would like to learn more about the solar neutrino problem, or the "hot" vs. "cold" dark matter theories, I provide a list of books for further reading at the end of this book. They are all suitable for general readership and these and many more details can be found in them.

Next question: What qualifies me to write such a book? I am not a professional scientist, though I do have a BS in physics. Far from disqualifying me, I contend that in some ways this uniquely qualifies me to write such a book. I don't have a professor's Ph.D ego to support. I don't have professional colleagues to impress. The only person I'm trying to impress, dear reader, is you, and then not with my stellar (pardon the pun) literary abilities, but with the awesomeness of the subject matter. I do have a certain amount of writing experience (see "About the Author" on the inside cover). I'm also a great believer that knowledge is no good unless you share it.

I gave careful thought to a title for the book. I wanted it to appeal to my target audience, and convey to a prospective reader what he/she would have gained when they were through reading. Astronomy For Dummies/Idiots had already been taken. More than that, I don't think you're dumb, and you certainly are not an idiot, otherwise you wouldn't have picked up this fine volume. You are intelligent and curious, but ignorant as regards this particular subject, the same way I'm ignorant of automotive repair. That will hopefully change after you read the following pages.

Astronomy Made simple? No. I don't believe any subject worthy of concentrated study is simple. Astronomy in a Nutshell? Not right either. The title of the class I originally wrote the book for was "Astronomy Basics," but the members of a certain writers' critique group that I belong to thought that title was too "blah." I think "Planets to Pulsars: A Citizen's Guide to The Universe" accurately reflects the subject matter and my target audience, but I let you be the final judge.

Even though I wrote the book as a complement to an astronomy class, as explained previously, you need not be enrolled in my class or any other to enjoy the book. All you need is an abiding curiosity about our universe. I have written the chapters, to the best of my ability, to stand on their own. If the chapter on "Searching For E.T." interests you the most, then by all means read it first. Nothing should be lost in reading the chapters out of order (in fact I even wrote them out of order). I arranged the chapters in the order I did because that's the order I taught the subjects in my class, and the idea of starting with the objects closest to us, the planets and our solar system, and working my way out held a sort of logical appeal. Subjects that didn't seem to fit in any particular chapter, but which may nonetheless be interesting to some readers have been included as appendices.

All the information in the book comes from secondary sources, but since it is not intended to be an academic work, as explained previously, I have forgone the use of formal footnotes or endnotes lest they frighten away some of my intended readers. Instead I have listed the sources I have used for the information in each chapter alphabetically by title at the end of that chapter.

I hope you enjoy and learn from these pages. If so, I will have accomplished my goal.

✳ ✦ ✳

CHAPTER 1

THE PLANETS: HEAVENS IN MOTION

Chances are that most of you reading this book, no matter your vocation or educational background, have been required at some point in your life to take a science class. Therefore, it's also a safe bet that, whether you remember it or not, most of you have also been exposed to one or more explanations of the "scientific method." The absolute best (simplest and most unpretentious) explanation of the scientific method I've ever heard came from Nobel Prize-winning physicist Richard Feynman. I never met Dr. Feynman, but several years after his death I watched a PBS NOVA program on his life. Included in the show was some old black and white footage of Dr. Feynman lecturing on the scientific method. With all due credit and respect to Dr. Feynman, I paraphrase it here.

The first step in discovering how any physical system works is to take a guess. As simple as that. How do you think the system works? Any guess will do, although you'll save some time by making educated guesses. I could guess that soap bubbles are round because my eyeballs are round, but that would be a rather silly guess. Besides which, why wouldn't everything else we see be round in that case? For the most part however, one guess is as good as another.

The second step is to work out the implications of your guess. Here is where a scientist separates the plausible guesses from fantasy. A good guess must be specific enough to allow the scientist to make predictions about the behavior of the system in question, which can then (hopefully) be tested in the laboratory. In other words, the scientist must be able to say, "If my guess about the way the system works is right, then if I do A to the system, B should be the result." The initial guess should allow the scientist to predict how the system will react to certain stimuli.

The third step is to test the implications worked out in step two. An experiment or experiments are performed under controlled conditions. That is, as much as is possible, the system is isolated from any outside influences that might affect the outcome of the experiment, then the stimulus A is applied and the result observed. If the result is the predicted behavior B then this is an indication that your initial guess was correct. Note, I did not say that the result was proof of your theory.

Proof is elusive in science. Anyone seeking concrete proof should go into politics, law, or better yet, mathematics. The scientific method is not a means of proof. All a scientist has is a greater or lesser degree of confidence that a theory is correct. Even such well-established theories as Newton's laws of gravity and motion are constantly under scrutiny and subject to change. Indeed, some recent evidence has caused scientists to question whether Newtonian gravity might not have to be modified under some circumstances.

If the result of the experiment in step three is not what was predicted the scientist proceeds to step four. First, let me say that in astronomy step three gets modified a bit. It is not practical to bring a neutron star, or nebula or any other large astronomical body into the laboratory. In astronomy, the testing phase of the scientific method is reduced to the test of further observation.

In step four, the scientist makes any necessary revisions or adjustments to his original guess suggested by the results of the experiment. In other words, if a stimulus A was introduced to the system of interest and instead of the anticipated result B, the result was in fact C, the scientist says, "How can I modify my original guess to make it more consistent with the result of my experiment?" Sometimes it may not be possible to modify the original guess enough. It may be necessary to come up with a completely new guess. At any rate, the process begins all over again. The implications of the revised guess are worked out and tested, and so on until you have sufficiently high confidence that your theory is the way the system truly works.

With this description of the scientific method in mind, my exercise for you is to put yourself in the shoes of an early, pre-telescopic astronomer and apply the method to what you see in the night sky. The first thing you'd probably notice is that the heavens are in a constant state of motion and change. The Sun rises and sets. The Moon rises and sets, and goes through phases. The seasons change. Even the tiny points of light in the night sky move over the course of an evening or months.

Almost from the first time humans looked at the night sky, they must have noticed that these tiny lights were of two types. One type, the stars, came in a variety of brightness and colors. Over the course of months, these early observers would have noticed that the rising and setting times of the stars changed. Certain stars were visible only at certain times of the year, but the position of the stars relative to one another did not change, at least as far the naked eye could tell (see Chapter 3). Man grouped the stars (sometimes with quite a bit of imagination) into objects familiar to him: animals, characters from myth, gods, and goddesses. We call these groupings "constellations." Man learned to use the rising and setting of the constellations as a guide for when to plant and harvest crops, when to perform civil and religious ceremonies and as aids to navigation.

The other lights in the night sky (there were five of them) were different from the stars in at least two important ways. First, and most immediately noticeable, they were usually, though not always, brighter than the stars. Second, these lights were not stationary. Over the course of weeks and months, they moved within the constellations and even from constellation to constellation. The Greeks gave these lights the name "planets" which means wanderers, in honor of their motion.

Early observers were curious. What mechanism drove the lights in the night sky? Were these points of light (particularly the planets) dim and nearby, or bright and very far away? How did the Earth and human beings fit into the scheme of things? What might have been a logical guess as to how the universe worked?

Certainly, the Earth could not be moving. A moving Earth was, in fact still is, contradicted by everyday experience, as can be demonstrated by a simple experiment. Hold this book, or any convenient object, at chest height and let it drop. Where does it land? At your feet, of course. If the Earth were spinning, then in the time it took for the book to fall, the Earth would have rotated beneath it. The book would not land at your feet, but some distance to the west. If the Earth were spinning, then everything that was not nailed down would go flying off into space. If the Earth were traveling through space, then we would experience continuous winds in the opposite direction of our motion. The Moon is always in the sky. If the Earth were traveling around the Sun, we would leave the Moon behind. Since we experience none of this, the Earth must be stationary. This was/is common sense.

The master of common sense in the 3rd century BC was the mathematician, philosopher Aristotle of Alexandria Egypt. Aristotle and his groupies were obsessed with the perfection of mathematical equations, and geometrical shapes. As far as the Aristotelians were concerned, the workings of the physical universe should be no less perfect.

The most perfect shape the Aristotelians could think of was the sphere. Even today astronomers, amateur and professional alike, refer to the "celestial sphere." Aristotle's theory was that the Sun, Moon, and the five naked-eye-visible planets (Mercury, Venus, Mars, Jupiter, and Saturn) rested on the surface of crystalline spheres, turning at constant rates centered on a stationary Earth. Beyond this was the sphere of the stars. All of these objects thus traced out circular orbits as the spheres rotated.

This answered the how of the matter, but being a good scientist Aristotle even came up with an answer for why the celestial bodies should behave this way. According to Aristotle, everything in the universe was made of five elements: earth, air, water, fire, and aether. Each of these elements had its own "natural" state of motion. Everything on Earth was made of one of the first four elements, or a combination thereof. The natural state of motion of water and earth was to fall, or move towards the center of the Earth (therefore the center of the universe) because like attracts like. The natural state of motion of air and fire was the opposite of water and earth. These two elements moved away from the Earth's center.

The Sun, Moon, planets, and stars were made of the aether. The aether was incorruptible, meaning it did not mix with any of the other elements. The aether was also perfect and immutable. It had existed and would exist forever and did not blemish. The natural state of motion of the aether was uniform circular motion, or motion in a circle at a constant speed. Hence the motion of the heavenly bodies in Aristotle's cosmos.

It is so easy for us with our high-tech scientific tools to look back and say, "How could Aristotle have been so stupid as to think that the Earth was stationary at the center of the universe, and that the Sun, Moon, planets, and stars moved in circles on the surface of crystalline spheres?" But, as I hope the preceding discussion has shown, he was making the best guess he could with the information available to him. As silly as it may seem to citizens of the twenty-first century, it was a completely valid scientific guess. Where it failed was in the test of further observation.

Every now and then, the planets perform acrobatics. Any careful observer will notice that on occasion a planet in its usual trek across the sky will stop and begin to travel in the opposite direction. After traveling in reverse for a few days or weeks the planet again throws on the brakes and starts moving in its usual direction. This celestial loop-the-loop is called "retrograde motion." How could the theory of planets resting on crystalline spheres turning at constant rates be modified to account for this?

If a few spheres were good, maybe more would be better. By nesting spheres within spheres like Russian dolls, each turning at a different constant rate with their axes of rotation tilted with respect to each other, it is possible to recreate the loop-the-loop motion of the planets (diagram 1-1,1-2). Yet problems remained.

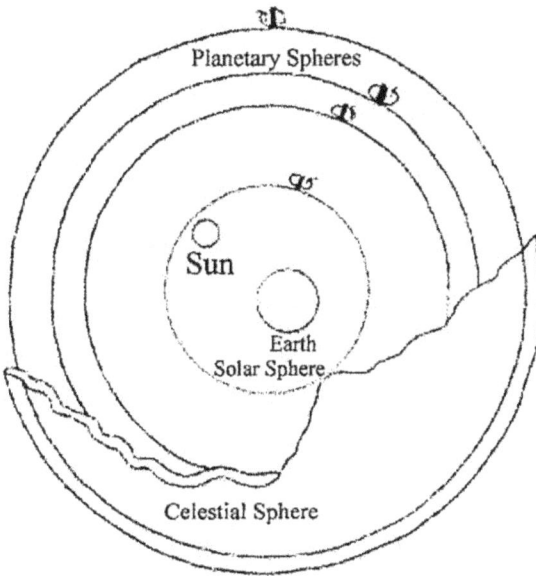

Diagram 1-1, 1-2

Aristotle's universe consisted of spheres nested within spheres, their axes and directions of rotations adjusted to approximate the observerd motions of the sun, moon, and starts across the sky. (Not to scale.)

Observers noticed that the planets were not always the same brightness. They were brighter sometimes and dimmer at others, as though they were moving closer to us and then farther away. How could this be if a planet rested on a crystalline sphere every point of which was equidistant from the stationary Earth? Furthermore, the motion of the planets through the constellations did not seem to be constant. They appeared to move faster sometimes, slower at others. This was definitely contradictory to the doctrine of uniform circular motion.

In the first century AD, Claudius Ptolemy did away with the crystalline spheres of Aristotle while retaining the notion of uniform circular motion and the other properties of the aether. In Ptolemy's model (diagram 1-3), the planets moved in small circles called "epicycles." The center of these small circles in turn moved in big circles centered not around the Earth, but an imaginary point called the "equant" offset from the center of the Earth. Motion of the planet around the epicycles and of the epicycles around the big circles, called deferents, was a constant, but the problems of the previous paragraph appeared to have been solved.

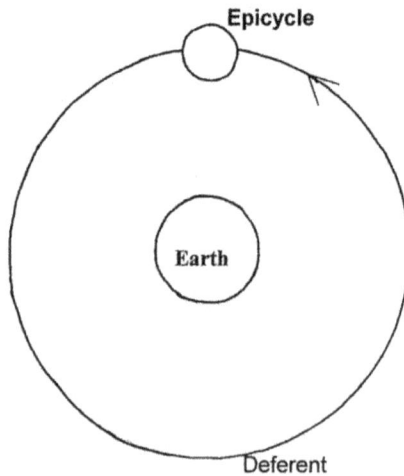

Epicycle

Earth

Deferent

Diagram 1-3

In the Ptolemic model of the solar system, the planets orbited the Earth on small circles (epicycles) the centers of which in turn move around the large circles called Deferent's. This complicated arrangement was needed to account for the "retrograde" motion of the planets.

The planets appeared brighter at times, as though closer to Earth, because they were closer when they were on the inside of the small circle. Likewise, when they were on the outer part of the small circle they were farther away and looked dimmer. The planets moved faster through the constellations at times because at certain times the motion of the planet around the little circle added to the motion of the big circle around the Earth (diagram 1-4). The planet appeared to move slower because at times the motion of the planet around the epicycle worked against, or cancelled to some extent, the motion of the big circle around the Earth.

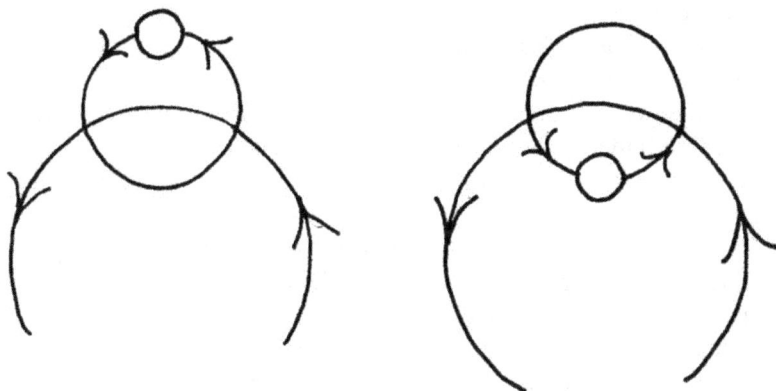

Diagram 1-4

On the outside of the epicycle the motion around the epicycle adds to the motion around the deferent, making the planet appear to move faster through the sky.

On the inside of the epicycle the motion is opposite the motion around the deferent and the planet appears to move more slowly.

With a total of forty circles in all, Ptolemy could predict the times of eclipses and conjunctions (close pairings of the planets), and the positions of the planets with fair accuracy months, even years in advance. It's difficult to know whether Ptolemy believed this was actually the way the solar system worked or if it was just an expedient for calculation, but if it ain't broke, why fix it? However, there is another, unwritten part of the scientific way of thinking that says whenever possible models of nature should be beautiful and simple. Ptolemy's cosmos may have had aspects of beauty with its interlocking circles, but with a total of forty of them, many must have found his theory decidedly lacking in simplicity. More than a few astronomers must have wondered if there wasn't a better way to account for the planets' motions.

There were murmurings of another theory, not necessarily new, but a radical, even heretical theory that the Earth was just another of the planets orbiting the Sun. It was heretical because the idea of a moving Earth was directly contradicted by the literal interpretation of scripture. In the Book of Joshua does it not say, "…Joshua spoke with the Lord, and he said in the presence of Israel, 'Stand still, O Sun, over Gibeon and Moon, you also over the valley of Aijalon.' And the Sun stood still…." How could the Sun stand still if it were not moving in the first place? To teach publicly that the Sun stood still and that the Earth moved around it was to invite imprisonment, torture, or death by church officials.

Nicholas Copernicus was not the inventor of the heliocentric (Sun-centered) theory, but his name is the one most commonly associated with it. It is possible that Copernicus may have read of the idea, or been introduced to it by one of his teachers. At any rate, Copernicus correctly dethroned the Earth from its central location and placed it as the third planet from the Sun. Unfortunately for Copernicus, he was still a slave to the idea of uniform, circular motion. He was forced, therefore, to retain the circles and epicycles of Ptolemy and the dreaded, imaginary equant. In fact, far from simplifying matters, Copernicus complicated them further. He ended up with eight more circles than Ptolemy. This must have disappointed Copernicus. For that reason and for fear of reprisal from ecclesiastical authorities, Copernicus waited to publish his theory until he was on his deathbed.

There were other objections to the Sun-centered model. If the Earth were just another planet orbiting the Sun, then over the course of months as the Earth traveled in its orbit the stars should exhibit a measurable parallax (see Appendix C). Parallax is fairly easy to demonstrate. Pick an object at the opposite end of a room (light switch, poster, whatever is handy) and stand facing it. Hold a pen in front of you at arm's-length. With the other hand cover first one eye and then the other and observe how the position of the pen changes with respect to the background object, appearing first on one side of the object then the other. This perceived change of position is called parallax. The fact that the stars did not show any visible parallax meant that either the Earth was stationary, or the stars were incredibly far away, an idea almost as hard to swallow as a moving Earth. It would take very accurate measurements to determine which was the case.

At the same time, increasingly accurate measurements were revealing further flaws in the Ptolemaic model. The undisputed king of pre-telescopic observation and measurement was the Danish astronomer Tycho Brahe. Tycho was not a Copernican, but his observations were causing him to increasingly lose faith in the Ptolemaic model. In 1572, Tycho observed a "nova," or "new star" in the constellation of Cassiopeia. He was reportedly so startled at this apparition that he stopped some passers-by traveling in a carriage to ask them if they saw it also. The nature of the aether meant that stars should be eternal. How could there be a new star?

The accuracy of Tycho's measurements also revealed a weakening of the Ptolemaic model's predictive power. In August of 1563, Tycho observed a conjunction (a close pairing in Earth's sky) of the planets Jupiter and Saturn. He noticed that the timing of the closest approach of the two planets was days off of the time predicted by Ptolemy's theory.

Tycho didn't want to throw the baby out with the bathwater. The idea of circular orbits was too appealing to just discard out of hand. He came up with a compromise. In the Tychonic model, Sun and Moon still orbited the Earth but the rest of the planets orbited the Sun.

Tycho needed to find some evidence in favor of his theory. He chose to try to measure the distance to the planet Mars. In the Tychonic system, Mars would come closer to the Earth than the Sun does. It also happens that Mars is the planet in which the loop-the-loop action discussed earlier is most pronounced.

Pre-telescopic instruments were limited to devices, which could measure the separation of objects in the sky, or their position on the dome of the heavens (essentially the latitude and longitude of an object on the celestial sphere). One such device was the "quadrant," so named because it took the form of a quarter circle with a sighting device attached to a pivot at the vertex (diagram 1-5). The user would look along the sighting device until he found the object of interest, then an assistant would read off the elevation of the object in degrees from the outside arc of the quadrant. The larger the quadrant, the greater the accuracy of the measurements, down to the limiting accuracy of the human eye.

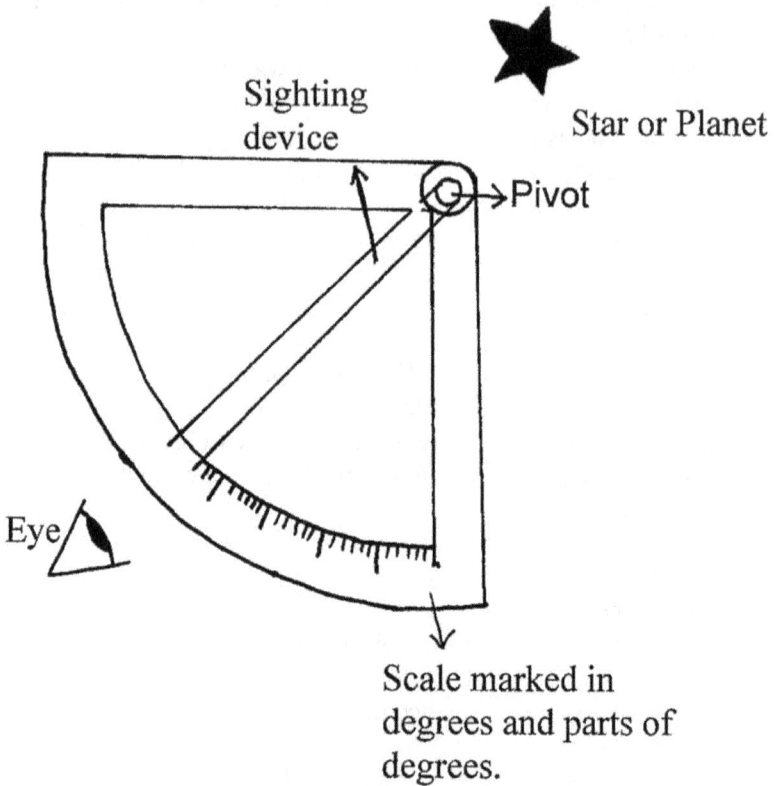

Sighting device

Star or Planet

Pivot

Eye

Scale marked in degrees and parts of degrees.

Diagram 1-5

A quadrant, appropriately named after the quarter circle it encompasses, was a pre-telescopic device for measuring the elevation of a star or planet. The observer would site along a straight edge and read the elevation off the scale.

Tycho used his great wealth to procure and build the largest, most accurate instruments of his day. For years when the viewing conditions were optimal Tycho made the most painstaking measurements of the position of the planet Mars. Try as he might though he could not make sense of his measurements. Tycho was a stunning observationalist, but his mathematical skills were not up to the task of unraveling the solar systems mysteries.

In the year 1600, Tycho took as his assistant a German astronomer and mathematician named Johannes Kepler. Tycho and

Kepler were the "odd couple" of astronomy. Tycho was an experimentalist. Kepler was a theoretician. Kepler was dirt poor. He made enough money to survive by making astrological charts for wealthy patrons. Tycho was wealthy as described previously. Tycho loved to throw lavish parties and be the center of attention. Kepler was so meek that as a teacher he had trouble holding the attention of his students.

Tycho parted with his Mars data grudgingly. This was a further source of tension between the two. Kepler went so far as to compare himself to Tycho's dog, being fed scraps of information, like a dog being fed table scraps, whenever the mood hit Tycho. It was only after Tycho's death and a battle with his relatives that Kepler finally got his hands on the bulk of Tycho's Mars observations.

At first, Kepler couldn't make sense of them either. He had his own preconceived notions of how things ought to work. For those interested in the details of how Kepler came to the conclusions he did, I refer the reader to the section on suggestions for further reading. Suffice it to say that after a few false starts in 1609 Kepler formulated the first two of what have been immortalized as Kepler's Laws of Planetary Motion.

The first law says that the planets do indeed orbit the Sun, but not in circular orbits. Rather the planets follow egg-shaped, mathematical curves called ellipses. Furthermore, the Sun is not at the center of the ellipse (an ellipse doesn't really have a center) but rather at one "focus" (ellipses have two internal points, or "foci"). The distances of the planets from the Sun and from Earth vary throughout their orbits. When the planets are closer to Earth they appear brighter, and vice-versa.

The second law says that the planets <u>Do Not</u> travel at constant speeds. What is constant, according to Kepler, is the area swept out by the planets in equal time intervals. In other words, if you were to tie one end of an imaginary rope to the center of the Sun and the other to a planet then set the planet in motion for a certain length of time, the rope would sweep out a certain area of space (see illustration 1-1). No matter where in its orbit you started the clock if you let exactly the same amount of time pass and measured the area, it would be the same. This is really a mathematical way of saying that a planet travels faster when it is closer to the Sun in its orbit, and slower when it's farther away.

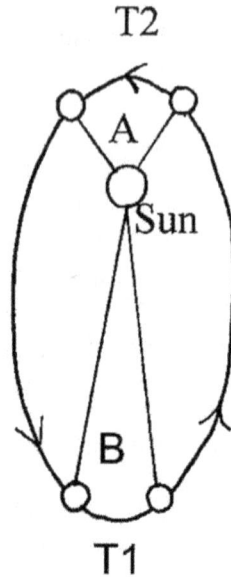

Illustration 1-1

A planet in its orbit sweeps out an area B in time T1 and an area A in T-2. Keplers second law of planetary motion says that if T1=T2, then A=B. This is really just a fancy way of saying planets move faster when they're closer to the Sun and more slowly when they are farther from the Sun.

It took Kepler ten more years to formulate his third and last law of motion, even though at first glance it may seem self-evident. The third law says that the amount of time it takes a planet to make one complete orbit of the Sun is proportional to its distance from the Sun. Jupiter takes longer to complete one orbit than Earth because Jupiter is farther from the Sun than Earth. The loop-the-loop motion that caused Aristotle and Ptolemy all those problems is an optical illusion. As the Earth, on the inside track as it were, overtakes Mars in its orbit, Mars looks as though it's traveling backwards (diagram 1-6).

Apparent path of Mars in the sky

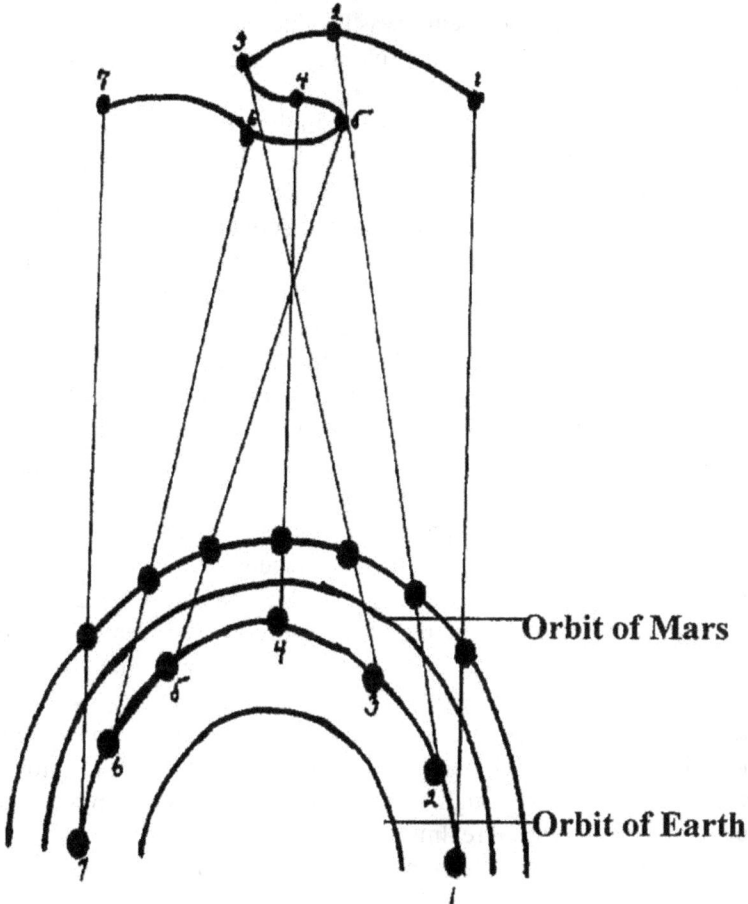

Diagram 1-6

Retrograde motion of Mars occurs when Earth overtakes the more slowly moving outer planet, making Mars appear to move backward in the sky.

At around the same time that Kepler was formulating his laws of planetary motion arguably the most famous astronomer of all time, Galileo Galilei, was using a new piece of technology called the telescope to add to the list of evidence against the Ptolemaic system. With his improvements in the design of the telescope, (see Chapter 5) he was able to see the phases of the planet Venus. Galileo also noticed a change in brightness and size of the planet as seen through his scope associated with the changes in phase. Venus was brighter and larger in its crescent phase than in the full phase. The Ptolemaic model of the solar system could not account for this.

When he turned his telescope at Jupiter, Galileo saw some points of light near the planet. These points of light, which Galileo called the "Medician Stars" after the Grand Duke of Tuscany, seemed to follow Jupiter across the sky. Galileo eventually discovered four of these bodies we now call the "Galilean moons" in his honor. Obviously not every body orbited the Sun, and if Jupiter could hold on to four moons as it orbited, certainly the Earth could hold onto just one.

But though the question of how the planets moved was solved, that still left the question of why they moved the way they do. What held the planets in their orbits? Why did the planets orbit in ellipses? Why didn't they go flying off into the vastness of space? The second and third laws of planetary motion seemed to suggest that the motive force lay in the Sun, but other than that Kepler didn't have a clue. The answer would come from one of the greatest Christmas presents the world would ever receive. His name was Isaac Newton, born December 25, 1642.

Newton did not invent the idea of gravity. The idea of a force that pulled objects to the ground when they were dropped had been around since the time of Aristotle. Some very brilliant minds worked on the theory of gravity including Christopher Wren, and Newton's arch nemesis Robert Hooke (Hooke even accused Newton of having stolen his work). Newton's breakthrough was in making the connection between the celestial and terrestrial realms.

The story of how Newton reached his epiphany regarding gravity probably has about as much truth in it as the beloved American story of George Washington and the cherry tree. As the story goes, Newton was visiting the family estate in Woolsthorpe during the summer of 1665 trying to escape the Black Death that was ravaging Cambridge at the time. One day he was sitting under an apple tree (the estate did have an orchard) watching the Moon hang lazily in the sky when he witnessed an apple fall to the ground. As he looked back and forth between the apple and the Moon one of the most basic questions a scientist can ask occurred to him, "What if?" What if, contrary to the philosophy of Aristotle, there were not two separate sets of rules for how bodies moved, one set for earthly bodies and another for Sun, Moon, planets and stars, but rather one set of rules for every body in the universe? What if the Moon was falling around the Earth and the planets falling around the Sun, with gravity keeping them from flying off into space like some invisible string.

Imagine a canon. Put some powder and a ball down the muzzle and light the fuse. BOOM! The ball comes flying out. Immediately gravity begins to pull on it. The forward velocity of the ball combined with the downward pull of gravity causes the ball to follow a curved trajectory called a parabola (diagram 1-7). Eventually the ball will hit the ground some distance away. Put a little more powder in the canon, the ball travels a little farther. The curve the ball follows is a little bigger (diagram 1-8). If you could fill the canon with enough powder and didn't have to worry about the effects of air resistance on the ball, the curve of the ball's trajectory would follow the curve of the Earth's surface. The ball would travel all the way around the world and hit you in the back of the head (diagram 1-9). It would have completed one orbit.

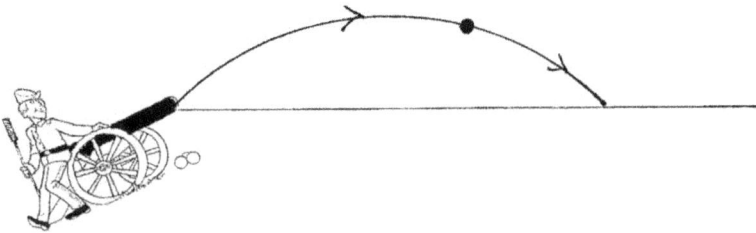

Diagram 1-7

When you fire a cannon, the ball travels a curved path called a parabola. After a distance, air resistance brings the forward motion of the ball to a halt.

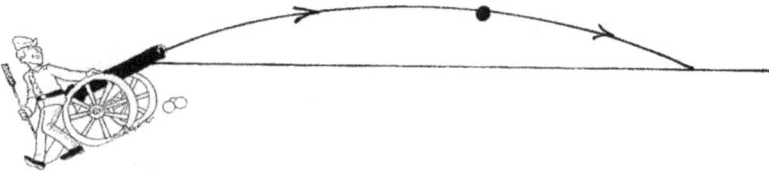

Diagram 1-8

Add a little more powder to the cannon, and the ball goes a little farther; and the trajectory is a little flatter.

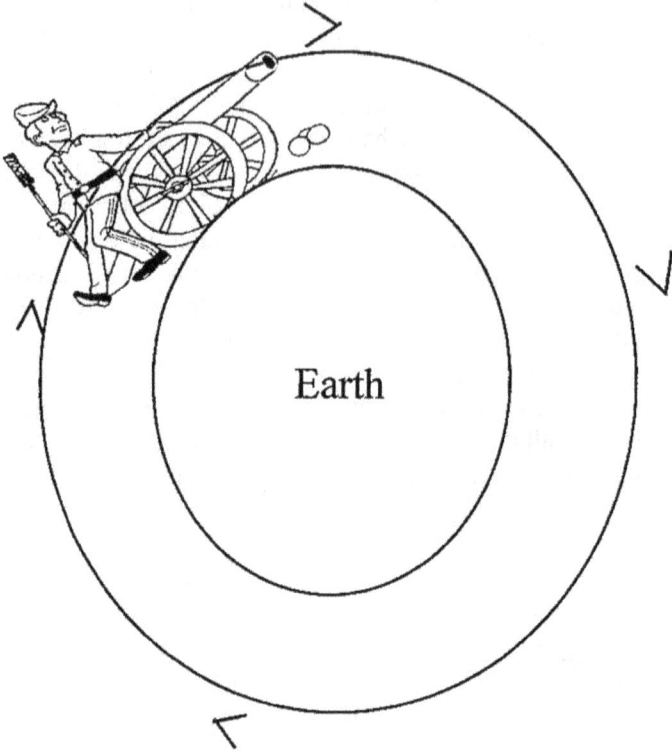

Diagram 1-9

If it were possible to put enough powder in the cannon, and if air resistance were nil or negligible, the ball would travel completely around the Earth and hit you in the head. It would have completed one "orbit."

Newton also gave us the equation that any high school student who takes a physical science class uses to calculate the gravitational force between two objects: $F=GMm/r^2$. The force is equal to a constant, immortalized as Newton's Gravitational Constant, times the mass of one of the objects times the mass of the other, divided by the distance separating them squared (multiplied by itself). Using this equation and his laws of motion along with his newly developed tool of Calculus, Newton was able to derive all of Kepler's laws of planetary motion from first principles.

Nevertheless, Newton was reluctant to publish his findings. Newton didn't handle criticism well and was no doubt afraid of recriminations from Hooke. It was only with the urging and financial backing of Edmund Halley of Halley's Comet fame that Newton published his Principia Mathematica that contained his laws of motion and equation of gravity in three volumes.

The third volume contained his work on gravity. The first established his laws of motion. The first law of motion is familiar to almost anyone, scientist or not. It says that bodies in motion tend to stay in motion in a straight line unless acted upon by a force. Likewise, bodies at rest tend to stay at rest unless acted upon by a force. This law of "inertia" is why a book dropped at shoulder height falls at your feet instead of some distance away. You, this book, and the lamp you're reading it by are all spinning with the Earth. We share the Earth's state of motion. When you release the book, it continues to spin with the Earth because it has mass and therefore inertia.

The second law says that the force on a body is equal to its mass times its acceleration. The third law says that for every action (force), there is an equal, in magnitude, but oppositely directed force produced. When I jump rope, which I don't often do, I push on the Earth, and the Earth pushes back on me. I travel into the air, but the Earth also moves (a tiny bit because of the Earth's much greater mass) in the opposite direction. As Jupiter orbits the Sun the Sun pulls on Jupiter, but Jupiter also pulls on the Sun. Astronomers use this effect all the time to search for planets around other stars (see chapter 6). But it was first used more than 100 years after Newton formulated his laws of motion to predict the existence of a new body in our solar system.

The discovery by Galileo of the moons of Jupiter suggested that there might be other bodies in our solar system as yet undiscovered. Where were they? For hundreds of years all astronomers had been aware of the five planets known to the ancients (Mercury, Venus, Mars, Jupiter, and Saturn). Then on March 31, 1781, an avid amateur astronomer named William Herschel saw something he did not recognize in the constellation of Gemini. Herschel was an amateur, but he knew the night sky well enough to know when something was out of place. What's more, this object wasn't point-like as a star would be. Herschel believed he could make out a disk shape through the eyepiece of his telescope.

Maybe the object was a comet and Herschel was seeing the coma (see chapter 2). The object didn't move like a comet though. It soon became apparent that Herschel had discovered a new planet. After some debate, it was decided to call the planet Uranus. Little did Herschel know that the discovery of Uranus was the beginning of a much bigger mystery.

As the planets orbit the Sun, they also feel the gravitational influence of all the other planets (diagrams 1-10, 1-11). The problem is that the equation that would take into account the gravitational affect of all the other planets on one particular planet cannot be solved exactly. Let me repeat this. The "multi-body problem" as it's called is not only difficult to solve IT'S IMPOSSIBLE. The mathematician Pierre Simon LaPlace came up with a way to approximate the solution, however.

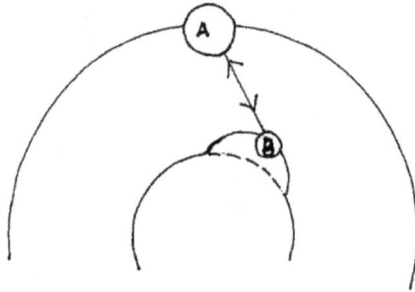

Diagram 1-10

Planet A pulls on planet B, speeding it up and pulling out of its normal orbital path.

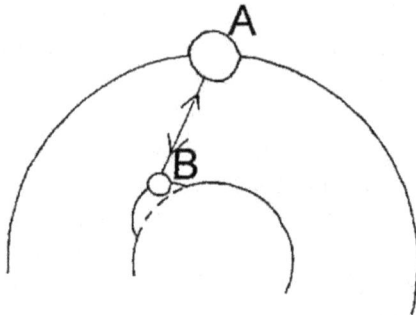

Diagram 1-11

Planet A pulls on Planet B, this time slowing it down. By analyzing effects like this on Uranus, Adams and LeVerier postulated the existence of Neptune.

In Laplace's formulation, the orbit of a planet after the gravitational influence of its counterparts has been taken into account, or the "perturbed orbit" of the planet, can be approximated by starting with the "true ellipse" for that planet (the orbit the planet would have if it were the only planet orbiting the Sun) plus some extra perturbation terms. How many perturbation terms? As many as you need to get to the desired accuracy. In most cases, three or four is more than sufficient.

Because the gravitational force is proportional to the mass of the perturbing body it's usually good enough for government work to include only Jupiter and Saturn in the perturbation calculations since they are the most massive of the planets (the ones most likely to have a noticeable effect). However, in the case of Uranus this was not good enough. Try as they might astronomers could not calculate an accurate orbit for Uranus. It would follow their predictions for a while and then errors would creep into the predicted position and grow until any table listing the future position of Uranus in the sky would be useless.

This was perplexing and frustrating for astronomers. So frustrating to British astronomer George Biddle Airey that he mentioned it in a report he made for the British Association For The Advancement of Science. A few years later a Cambridge mathematics student named John Couche Adams came across Airey's report and its mention of the problem in the school library. He set himself the ambitious problem of solving the riddle of Uranus's wayward orbit.

He started by redoing the perturbation calculations using revised estimates of the masses of Jupiter and Saturn and even adding a few more terms just to be safe. It didn't work. Uranus still stubbornly refused to comply with the calculations.

In a moment akin to Newton's, "what if" moment of a century before, Adams wondered if it might be possible to account for the discrepancies in Uranus's orbit by positing the existence of an as yet unseen planet beyond Uranus's orbit, pulling the planet this way and that as it went by. The problem then was to work backwards in essence. Adams had to deduce the orbit of the unseen planet based on its influence on Uranus so that people with telescopes could point them in the right direction to find this planet.

Because Adams did not know either the mass or the distance of the perturbing planet from Uranus, (both are variables in the equation for gravity) he had to take an educated guess at the unseen object's mass. The calculations were made even more difficult by the fact that there were no electronic computers. The good news was that all the other planets orbited the Sun in an imaginary plane called the "ecliptic" (see chapter 2). So assuming the undiscovered planet played by the rules Adams only needed to calculate one number (essentially the planet's longitude on the celestial sphere) in order for astronomers to be able to find it.

As daunting as Adams's calculations were, the really hard part was going to be convincing anyone with a telescope to take his work seriously. Then, as now, telescope time was precious. Observatory schedules were determined months in advance. Taking into account the affect of known bodies on other known bodies was one thing. Finding an observatory director willing to upset his observing schedule based on speculative calculations made by a relatively unknown mathematician was next to impossible. It didn't help matters for Adams that a French astronomer, Jean Joseph LeVerier was working on the same problem.

Working completely independently of one another and using different approaches to the problem Adams and LeVerier calculated exactly the same position for the planet. In the end LeVerier, with his slightly higher standing in the scientific community was able to convince Johann Galle and Heinrich D'Arrest to use the Fraunhofer telescope at Berlin Observatory to make the search. Galle and D'Arrest checked the positions of objects in the telescope's field of view with the position of known objects in a catalog until they came across a point of light where none had been previously recorded. The new planet was exactly where Adams and LeVerier said it would be.

The ensuing debate over who should get credit for the discovery and what the new planet should be named resembled the mudslinging of a modern political campaign. One French newspaper even went so far to publish a cartoon showing Adams using his telescope to peer across the English Channel into LeVerier's study so he could copy his notes. Despite the name calling by their supporters Adams and LeVerier became good friends. The planet was eventually named Neptune.

It took almost 2000 years to overcome the notion of planets revolving around a stationary Earth in uniform, circular motion. In the years since Newton watched the apple fall from the tree (if indeed he ever did) NASA scientists and engineers have learned how to use the gravity of the planets to change the course of spacecraft without using precious fuel. Centuries after the motion of the planets presented a mystery to earlier astronomers the heavens are in motion in ways that they would never have imagined.

<u>Sources</u>

<u>Coming of Age In The Milky Way</u>, Timothy Ferris, William Morrow & Company, Inc., New York, NY., 1988.

<u>Cosmos</u>, Carl Sagan, Random House, New York, NY., 1980.

<u>Five Equations That Changed The World</u>, Michael Guillen, MJF Books, New York, NY., 1995.

<u>The Neptune File</u>, Tom Standage, Berkley Publishing Group, New York, NY., 2000.

<u>Tycho And Kepler</u>, Kitty Ferguson, Walker & Company, New York, NY., 2002

<u>The View From Planet Earth</u>, Vincent Cronin, Quill, New York, NY., 1988.

✳ ◆ ✳

CHAPTER 2

COMETS, ASTEROIDS, AND METEORS

Travel with me in the time ship that is your imagination, into the past 4.5 billion years, give or take a few million. Our destination is a cloud of gas and dust, not unlike any one of a number of nebulae that you can see through a telescope from a suburban backyard on any clear night.

The gas and dust in the cloud are not stationary. The particles bump into one another randomly, but the cloud is also rotating. A physicist would say that the cloud has an overall component of angular momentum. Then something, we will likely never know exactly what, causes the cloud to start to collapse and the matter to collect in a central location. As more material gathers in this location, the gravitational pull becomes larger, pulling in more matter. As the matter becomes more compact, the rotation rate increases, like an ice skater pulling in his/her arms during a spin.

The force associated with this rotation causes a disk of material to form around the equatorial region of what would become our Sun. Over time, the particles in this disk collide and stick together, forming larger particles. These larger particles collide, and so forth, eventually becoming the planets. The planets of our solar system still orbit in the plane of this original disk of material (with the exception of Pluto, see Appendix A). Astronomers call the plane the "ecliptic."

The outer gaseous planets probably formed first. Some computer modeling even suggests that without the periodic pull of Jupiter to stir the pot, as it were, the inner planets might never have formed. There might still just be a disk of rocky debris circling the Sun. In some sense, we may owe our existence to Jupiter. Not all the matter in the disk, however, went into forming the planets. There was material left over. These leftovers come in two varieties: comets and asteroids. Together these objects represent the most pristine material in our solar system.

Comets are amalgams of ice, rock and dust. Most popular science books use the term "dirty snowballs" to describe comets. They consist of a solid, or mostly solid nucleus, with a layer of ice mixed with dust covering this. The ice is a combination of plain water ice and "dry" or carbon dioxide ice. As the comet gets warmed by the Sun the ice begins to sublimate, or go directly from the solid phase to the gaseous phase. Some of the gas forms an atmosphere around the nucleus called the "coma."

As the comet gets nearer the Sun, the process of sublimation continues and the escaping gases pick up dust from the surface of the comet. Light reflecting off the dust produces the (sometimes) magnificent tails that most people associate with spectacular comets. It is possible for comets to have more than one tail. Charged particles from the Sun can ionize, or strip electrons off gases from the coma causing them to glow like a neon sign. The result is a short, blue-green tail called an "ion tail." These tails always point directly opposite the Sun, and so as a comet rounds the Sun and heads back into the solar system it travels tail first.

As comets travel through the inner solar system, they leave behind a trail of debris. Most of the particles are no bigger than a grain of sand, but when Earth passes through the debris, these particles enter our atmosphere at very high speed. The particles compress the atmosphere in front of them causing tremendous heat which then vaporizes them producing the streak of light we call a "meteor."

Although most people use the terms meteoroid, meteor and meteorite interchangeably, they are different. A meteoroid is an object in space. A meteor is the streak of light we see in the night sky. A meteorite is a meteoroid that has survived its trip through Earth's atmosphere to land on the surface.

Meteor showers are best observed in the early morning hours for two reasons: (1) the point in the sky from which the meteors appear to originate, called the "radiant," is higher in the sky in the morning, (2) the Earth's rotation carries the morning side into the stream of particles, thus increasing the speed with which they enter the atmosphere, making them burn more brightly. Sporadic, or random meteors can be seen at any time of year at a rate of around ten per hour.

The Earth can also pass through the trail of debris left by a particular comet year after year. Due to an illusion of perspective, if you trace the path of meteors from such a periodic shower backward, they will appear to converge at one point in the sky called the "radiant." These yearly showers get their names from the constellation, which contains the radiant. The Perseids of August appear to come from the constellation of Perseus. The Leonids of November appear to come from Leo the Lion. Table 2-1 provides a list of some of the major recurring showers, their hourly rates, and the comets that spawn them.

Table 2-1
Annual Meteor Showers

Shower	Month	radiant	Rate (per hr.)	comet of origin
Lyrids	April	Lyra	12	Thatcher 1861
Perseids	August	Perseus	60	Swift-Tuttle
Draconids	October	Draco	variable	Giacobinni-Zinner
Orionids	October	Orion	25	Halley
Taurids	November	Taurus	12	Encke
Leonids	November	Leo	highly variable	Temple-Tuttle
Geminids	December	Gemini	60	Phaeton (asteroid)

Most comets travel in paths that take them once through the solar system and then back out into the outer reaches of space never to be seen again. A few are trapped by the gravitational pull of the Sun and planets into orbits that bring them back to our neck of the woods every few years. Even today astronomers, professional and amateur alike, eagerly await the appearance of a bright comet. Many ancient observers, however, viewed them as evil omens.

Comets have presaged their share of disasters. A comet was seen over Italy in 79 A.D., just a few weeks before the eruption of Vesuvius that buried Pompeii and Herculaneum. Another comet seemingly foretold the coronation (800 A.D.) and death (814 A.D.) of Charlemagne. Some even believe that a comet appeared at the destruction of Sodom and Gomorrah. In 1911 the Earth was slated to pass through the tail of Halley's Comet. Earlier that year scientists had found evidence of deadly cyanogen gas in the tail of the comet. Various hucksters and fear mongers made small fortunes selling gas masks to a fearful public.

Early philosophers thought comets and meteors were atmospheric phenomena. This is where we get the name "meteorologist" for the person that gives the weather report on the evening news. Meteorites were called "thunderstones" and explained as the consolidation of particles in clouds caused by the heat of lightning. By the late 1700s, however, evidence of the extraterrestrial origin of meteorites was beginning to accumulate. In 1802, English chemist Charles Howard analyzed the composition of four meteorites from widely separated falls and found them to be identical, indicating a single cosmic origin for all of them.

It was the astronomer Tycho Brahe (see Chapter 1) who put the theory of the atmospheric origins of comets to rest once and for all. By observing the comet of 1577 and measuring its parallax (see Chapter 1 or Appendix C), he was able to determine its distance to be beyond the atmosphere of Earth and even beyond the orbit of the Moon. Although several of Tycho's contemporaries challenged his observations he was vindicated in the fullness of time.

Modern astronomers believe that comets reside in two distinct communities. The Oort cloud is a roughly spherical collection of objects extending from beyond the orbit of Pluto out to about 10,000 times the Earth-Sun distance (called an AU or "astronomical unit"). The Kuiper Belt starts around the orbit of Neptune and extends to about 50 AU.

The other class of leftovers, the asteroids, are rocky-metallic bodies, most of which are concentrated in the region between the planets Mars and Jupiter. Once thought to be the result of a cosmic collision of planet-sized objects, the current view is that the gravitational pull of Jupiter never allowed the material in the "asteroid belt" to form planets. There are gaps in the belt where no asteroids are allowed to reside. These spaces are called "Kirkwood gaps" after Daniel Kirkwood who discovered that any asteroids in the gaps would have orbital periods that were simple multiples of Jupiter's period. The periodic tug of Jupiter's gravity would therefore pull these bodies into different orbits, or out of the solar system entirely.

The very first of these objects was actually discovered by accident. In the aftermath of the discovery of Uranus by William Herschel (see Chapter 1), a wave of "new-planet fever" washed over the astronomical community. A German by the name of Franz Xavier von Zach was particularly hard hit. He traveled to meet Herschel, consult him on his methods, and examine his telescopes.

Von Zach's idea was to organize a group of 24 like-minded astronomers into a planet-finding organization which he planned to call

(I kid you not) the Celestial Police. Each member of the Police would be responsible for observing his own area of the Zodiac. Von Zach even established a professional journal so that members of the Police could communicate their findings.

In the autumn of 1800, von Zach met with five prospective members in Hanover, but before he put the plan into full swing, astronomer Guisseppi Piazzi beat von Zach to the punch. Piazzi spent several years working on a new catalog of the positions of 6,000 stars. In January 1801, he recorded the position of a "star" in the constellation of Taurus. When he observed the same area of sky a few days later, the "star" had moved. A few days later, it moved some more. This first main-belt asteroid was named Ceres after the goddess of the harvest and patron goddess of Sicily, home to Piazzi's royal supporter, King Ferdinand.

Then in March 1802, another prospective Police member, Heinrich Wilhelm Olbers, discovered another object in the same region of space between the orbits of Mars and Jupiter. When William Herschel invented a device to measure the size of these objects, he discovered that they were much too small to be planets. They looked star-like, and Herschel named them asteroids from the Latin "astrum" for star.

As technology improved and more astronomers began looking for them, discoveries of asteroids abounded. Even modern telescopic observations, however, reveal only so much about these enigmatic bodies. Most of what we know about them comes from pieces of them that have found their way to Earth. Indeed, by looking at the spectrum of reflected light from asteroids and comparing the indicated elemental composition of meteorite specimens, we can trace some of these objects back to their parent asteroids.

Meteorites fit into three basic categories called stony, stony-iron, and iron. These are much like the sedimentary, igneous, and metamorphic classification of terrestrial rocks. Stony meteorites are further subdivided into chondrites and achondrites. The chondrites get their name from tiny spherical inclusions of the minerals olivine and pyroxene called "chondrules." These vary in size, but a typical diameter would be about 1mm. It's not clear how chondrules form.

Achondrites are the result of melting and recrystallization of chondritic parent bodies. They are the igneous rocks of other worlds. The most famous of the achondrites are pieces of the planet Mars. How can we be sure? Robotic probes from Earth have sampled the composition of the Martian atmosphere. When samples of these meteorites are vaporized and analyzed, they are found to contain exactly the same gases in the same proportions as the atmosphere of Mars.

These pieces of Mars, liberated by the energy of some cataclysmic impact, go by the name of SNC achondrites. The most famous of these has the scientific, if un-poetic, name of ALH 84001. The ALH stands for the Allan Hills ice field of Antarctica where it was found. What makes it so famous is that inside this chunk of Mars, researchers found traces of organic molecules and curious worm-like, tubular structures. The debate continues as to whether these strange features are fossil evidence of Martian life.

The chondrites and achondrites are the crusts of other worlds. Their presence on Earth is evidence of the violence of the solar system. But as violent as the collisions that produced the chondrites and achondrites are, they are mere cosmic fender-benders. The stony-irons are evidence of much more violent impacts. Stony-iron meteorites are the mantles of solar system bodies, and iron meteorites are their cores.

Imagine an impact large enough to shake the Earth to its very core, and you will have some idea of the force of the collisions that produced the iron meteorites. The irons are separated into two groups based on their crystalline structure. This structure is very apparent in the Octahedrite group. When the polished face of one of these is brushed with dilute nitric acid, the acid eats unevenly at the different nickel-iron alloys that make up the meteorite. The result is the crisscross design known as Widmanstatten Patterns (see picture), named after scientist Alois Widmanstatten.

But, we don't need meteorite specimens to tell us that the solar system is a dangerous place. We have evidence in our own backyard. United States residents need only take a short detour (about 40 miles) east of Flagstaff, Arizona on I-40 to see some of this evidence, called Meteor Crater. Nestled in the otherwise bleak desert, Meteor Crater is proof that Earth is part of a cosmic shooting gallery. The crater is the result of a collision that happened around 20,000-50,000 years ago between Earth and an iron meteorite about 80 feet in diameter, weighing 63,000 tons. It struck the surface at about a 30 degree angle, traveling at almost ten miles per second.

The impact created a shock wave that moved downward through the Earth and up through the meteorite. The meteorite and the rock beneath it were compressed to less than half of their original volume. The rock was superheated, and the pressure was released in an explosion equivalent to 1.7 megatons of TNT. The resulting crater was 1,200 feet deep and almost half a mile wide.

If an impact of this magnitude were to occur in a modern city, it would be devastating. The energy released by the atomic bomb dropped on Hiroshima was a fraction of the energy released in the making of Meteor Crater. However, the human race would survive (it did survive). Not all species in the history of our planet have been as lucky. Evidence of one such event was discovered by a father and son scientific team looking for the answer to an entirely different riddle.

It's impossible to tell just by looking at a layer of sediment how long it took for it to be deposited. A thick deposit may have been laid down in hours by a flash flood. A thin layer may have taken hundreds of years to form. How to tell, then? Geologist Walter Alvarez hypothesized that the answer might come from the sky.

The surface of the Earth is showered with a steady rain of meteoric dust. This dust is in your hair, on crops, on your car. The dust contains the rare Earth element iridium. If he knew the rate at which the dust (i.e. iridium) was deposited on the surface of the Earth (say x grams per square centimeter per year) then by just measuring the iridium in a sedimentary sample, it would be possible to determine how long it had taken for that particular sedimentary layer to be deposited.

In the late 1970s, Dr. Alvarez was examining rock layers near the Italian town of Gubio, when he came across a very curious, red clay layer at the boundary between the Cretaceous and Tertiary periods of geologic time (called the K-T Boundary by geologists). This boundary is significant to geologists and paleontologists because it marks the end of the reign of the dinosaurs. Say what you will about the dinosaurs (dim-witted, cold-blooded, etc.), they were the dominant species on Earth for over 100 million years, Humans have been around for a tiny fraction of that time.

The red clay layer turned out to contain a tremendous excess of iridium, about 10,000 times what was expected. Where did it all come from? Walter consulted his physicist father Luis. Luis suspected that the answer was a collision between the Earth and an asteroid or comet almost six miles wide. Subsequently, similar layers were found at K-T boundary layers all over the world. But where was the crater?

Unlike Mars, or the Moon, the Earth is a dynamic world. Plates grind against and over one another, volcanoes erupt and cover previous surfaces, so it wasn't too surprising that the crater left by the impact responsible for the dinosaurs' destruction wasn't visible. Until it was found, however, critics of the impact theory could still point to the crater's absence as evidence against an impact.

By one of those remarkable coincidences of fate that seem to occur from time-to-time in the sciences, at about the same time that Walter Alvarez was examining the Gubio clay, Pemex, Mexico's national oil company, was conducting an airborne magnetic survey of the Yucatan Peninsula. A plane towing a magnetometer traced a grid, crossing the Yucatan coast to the Gulf of Mexico, looking for oil-bearing sedimentary beds.

Each day, geophysicist Glen Penfield pieced together the strip charts produced the day before. Slowly, a pattern began to appear in the form of a semicircular arc opening to the south. Penfield's interest was piqued, and he decided to do some further checking. He managed to find a gravity survey of the same area, also made by Pemex in the 1960s. Gravity surveys use devices that measure gravitational pull to look for differences in density of underlying rock layers. It showed a similar arc, this time opening to the north. The two features fit like pieces of a jigsaw puzzle to reveal an ancient crater 100 miles wide, centered on the town of Chiculub.

And yet, the critics still weren't satisfied. The iridium in the clay might have been the result of a major volcanic eruption. Penfield then found a core sample taken from just outside the crater rim. The sample contained "shocked quartz." Shocked quartz is ordinary quartz that has been subjected to tremendous extremes of pressure and temperature (150 tons/square inch and 1,000s of degrees Fahrenheit). Shocked quartz is **only** found at known impact sites such as Meteor Crater. The critics finally fell silent.

How frequently does this kind of species-eradicating collision occur? A group of researchers from Princeton claimed to have found evidence in the geologic record of a regular pattern of mass extinctions at 26-million-year intervals. They hypothesized that the culprit was an unseen Brown Dwarf companion to the Sun with an orbital period of 26 million years. Once each orbit, this unseen star, named "Nemesis," would disturb the Oort cloud of comets sending several Oort bodies on trajectories to the inner solar system.

No evidence of Nemesis has ever been found. Some scientists even doubt the reliability of the 26-million-year cycle. However, one thing is certain. Large bodies have hit the Earth in the past. It will happen again. Given that, what can we do to make sure we as a species don't go the way of the dinosaurs?

The first step would seem to be detection of the bodies, comets and asteroids that cross the Earth's orbit and therefore have the potential to collide with us. Before the advent of modern technology, this was an arduous task carried out by dedicated observers with an intimate familiarity with the night sky. Because asteroids reflect so little of the light that reaches them from the Sun, astronomers rely on their motion against the backdrop of the stars to give them away. The development of astrophotography simplified the search somewhat. Two photos of the same area of sky, taken days apart, were put in a device called a "blink comparator" (see Appendix A). By alternately shining light through or on one and then the other of the plates in rapid succession, the observer could see any object that was moving.

Today, the same sort of thing is accomplished using arrays of computer controlled telescopes and the latest in digital cameras. The blinking is done by computer software. This frees human astronomers to use their hard-won telescope time for more important research. One such computerized array is the LINEAR (Lincoln laboratory Near Earth Asteroid Research) project funded by the U.S. Air Force. Another such project is called NEAT (Near Earth Asteroid Tracking), run jointly from Palomar Observatory and the Maui Space Surveillance Site in Hawaii.

Even with all our resources focused on the problem, we can't expect to find all the potential impactors. A week or so before I wrote this sentence for the first time, a several-mile-wide body came within a few Earth-radii of our planet and wasn't discovered until it had already passed. But what if we did have advanced notice of an impact? What could we do?

That depends largely on how much advanced notice we had. Collision is a matter of two objects being in the same place at the same time. This is significant because given enough advance notice, a small nudge to the trajectory of the impactor will, over time, accumulate so that the body misses Earth. Contrary to all the Hollywood apocalyptic imagery, shooting an asteroid with a nuclear tipped missile would not be a good idea. By breaking the asteroid to pieces all that would be accomplished is to spread the devastation over a larger area. A better idea would be to explode the warhead near the asteroid using the energy released to push it out of the way without fracturing it.

Another option, given enough time, would be to land spacecraft on the asteroid and use rockets to push it out of Earth's way. Most asteroids are spinning, so the rocket firings would have to be carefully timed so as not to make the situation worse instead of better. Also, too much force applied in the wrong place, or force applied unevenly, might put stress on a fracture and break the rock into pieces just like the strike of a nuclear weapon.

Light exerts a pressure. If we were able to deploy a large enough, reflective surface on an asteroid, over time the pressure of sunlight might push it out of our way. Or we might be able to heat certain spots on a comet's surface, releasing gases that would move the object enough so that it would miss us. With decades of advanced notice, it might be possible to let gravity do the work. A spaceship could be sent to hover near the asteroid applying thrust of the correct amount, in the right direction, using the gravitational pull of its own mass to draw the body off course. The spaceship would in effect act as a cosmic tugboat.

Comets and meteor showers are awesome spectacles, but they are also reminders that we live in a cosmic shooting gallery. The earth will be struck again. Should we be worried? Yes. Should anyone lose any sleep over the prospect? No.

Sources

Astronomy For Dummies, Stephen P. Maran, Wiley Publishing Inc., Hoboken, NJ, 2005

Comets: The Swords of Heaven, David Ritchie, New American Library, New York, NY, 1985

Planet Earth, Jonathan Weiner, Bantam Books Inc., New York, NY, 1986

Rocks From Space, O. Richard Norton, Mountain Press Publishing Company, Missoula MT, 1994

When Life Nearly Died, Michael J. Benton, Thames & Hudson Ltd., London, UK, 2003

✳ ◆ ✳

CHAPTER 3

STARS, CLUSTERS, AND GALAXIES

In Chapter 1, I argued that it must have been fairly obvious to early sky watchers that there were two types of light in the night sky. The planets, stood out because they were generally brighter, and they also moved with respect to the other kind of light in the sky. But what about the other lights, the stars? What were they? Were they even physical objects at all?

Some of the earliest philosophers claimed that the stars were nothing more than holes in the dome of the sky, allowing the brilliant light of Heaven to shine through. In the early 1600s, however, when Galileo turned his telescope skyward, he saw many more of these lights than were visible to the unaided eye. Much later, Edmond Halley, namesake of the most famous comet in history (with the possible exception of Shoemaker-Levy 9 which crashed spectacularly into Jupiter), was commissioned by the Royal Society to produce a star catalog using earlier observations by John Flamsteed. Halley compared Flamsteed's data with measurements of the positions of the same stars made in the second century BC by Hipparchas and found that some of the stars had moved relative to each other in the intervening time. Clearly the stars were not holes in the dome of the sky. Neither were they fixed as though mounted on some celestial sphere.

As early as the sixth century BC, Anaxagoras postulated that the stars were incandescent stones. Some of these appeared dimmer, he claimed, because they were farther away. With his telescope in the

1600s, Galileo noticed dark spots on the Sun's surface. Might the Sun be a giant lump of coal, and sunspots nothing more than a window to our star's dark surface? Whatever the Sun (and by implication the other stars) was made of, whatever process fanned its flame, it must have been shining for as long as the Earth had been around, or at least as long as Man had been around to bask in its glow. But how long was that?

Until the advent of modern archeology and dating techniques, the leading source of information for those wanting to calculate the age of the Earth and Man's reign on it was the Bible. By adding up all the begets in the Bible and assigning an average age to each generation, the Reverend James Ussher calculated in the seventeenth century that the Earth was about 5,000 years old. A long time indeed for a lump of coal to burn, but perhaps possible. As it turns out, Ussher's estimate of the Earth's age was far short of the mark.

With the discovery of, and investigation into, the nature of radioactivity in the late nineteenth and early twentieth century's by Henri Becquerel and the Curies, a new method of dating the planet and objects on it became available. Atoms are made of protons, neutrons and electrons. Protons have a positive charge, electrons a negative one. Neutrons are electrically neutral. Most atoms have an equal number of all three. Carbon, for example, has a nucleus made of twelve neutrons and twelve protons, surrounded by a cloud of twelve electrons. But, as with everything, there are exceptions to this rule.

Atoms with differing numbers of protons and electrons, that is atoms with a net electric charge, are called ions. Atoms in which there are more neutrons than protons are called isotopes. Some isotopes are unstable. They want to spontaneously decay into more stable isotopes of the same element, or other elements entirely. Each unstable or "parent" isotope has its own characteristic half-life or time it takes for half of the original sample to decay into the more stable or "daughter" isotope. If half of a sample has decayed, then one half-life has passed. It will take another half-life for half of the remaining sample to decay, i.e. for one fourth of the original sample to remain, and so on.

All living things, plants, people, even dogs, take in carbon-14 from the environment. Upon the death of the organism, this carbon-14 begins to decay into the stable isotope nitrogen-14. By measuring the relative amounts of the two isotopes in a piece of wood, for example, scientists can determine how long ago the tree the wood came from died. Carbon-14 has a half-life of about 5,700 years. Fine for dating once-living things, but what about the Earth itself?

Potassium-40, uranium-235, and uranium-238 are all found in various mineral samples. All have half-lives of several billion years. Using these isotopes to date Earth rocks and Moon rocks yields dates of up to 3-4 billion years. It would take a tremendously large lump of coal to burn that long. But perhaps radioactivity itself held the answer. Maybe the same process that made it possible to date rock samples provided the power source for the stars.

Anyone who is able to read a book or watch TV thanks to a nearby atomic power plant knows that radioactive decay (fission) produces a tremendous amount of heat. Scientists think that a major part of the heat that drives volcanoes and plate tectonics here on Earth is the result of the decay of uranium in the Earth's core. And with a half-life of billions of years, a sufficiently large blob of uranium would still be producing energy. It also means, however, that the Sun's energy output would have decreased by half since the solar system's formation. We find no evidence of this.

It turns out that the Sun's energy, indeed that of all stars, is produced by a nuclear reaction, but not the decay of heavier elements into lighter ones, rather the melding of lighter atoms into heavier ones, or fusion. In order for the fusion process to be initiated, a certain threshold of temperature and confinement of the reactants must take place. The atoms must collide energetically (the temperature must be high), and the atoms must be confined to a small enough space so that they don't travel far before encountering or colliding with another atom. A physicist would say the "mean free path" must be small. The temperature at the center of the Sun is about 29 million degrees F. Nuclear fusion reactors on Earth must operate at much higher temperatures because it's impossible for engineers to compress the reactant gases as much as they are in the Sun.

A star is born when a tremendous cloud of mostly hydrogen gas collapses, becoming hot enough and compressed enough under the weight of the surrounding material that fusion begins, turning two hydrogen atoms into one atom of helium. The resulting helium atom is slightly lighter than the two hydrogen atoms that produced it. The tiny difference in mass of the reactants and the product is released as energy via Einstein's famous equation $E=mc^2$. Not every collision between hydrogen atoms produces helium. Only one in every 10 billion trillion collisions occurs at just the right angle with enough energy to produce a fusion. But the incredible amount of energy released and the overwhelming amount of hydrogen in a typical star means that it will "burn" for hundreds of millions if not billions of years. Eventually a balance is reached between the outward pressure produced by the fusion reaction and the force of gravity trying to compress the star, and it enters a comfortable middle age period of its life called the main sequence.

How long this period lasts ultimately depends on the star's mass. Like people, stars come in all shapes and sizes. Also like people, ultra massive stars live dramatically shortened lives. Stars with several times the mass of our Sun burn blue-white hot in order to keep gravity at bay. As a result, their fuel will only last several hundred million years before gravity wins the battle.

Medium-sized stars, like our Sun, burn yellow-orange and do so happily for billions of years. Our Sun should go on fusing hydrogen into helium contentedly for another 5 billion years before it begins to feel its age. Less massive stars burn a dull red and live twice as long as our Sun before passing into docile, doddering old age.

The Sun's core actually consists of two parts (see diagram 3-1). The outer core extends some 108,000 miles from the center. Moving outward from the core, the temperature decreases as you might expect. The convective zone starts about 307,000 miles from the center. Here hot gases rise to cool and fall back again like boiling water in a kettle. From the convection zone to the photosphere, the light producing layer, the temperature drops from about 4 million degrees F to some 9,900 degrees F. Then a strange thing happens. The temperature in the outermost layers of the Sun's atmosphere increases. This would be like moving a thermometer away from the center of a campfire, watching the temperature drop until you got several feet from the heart of the fire and then suddenly watching it shoot up again.

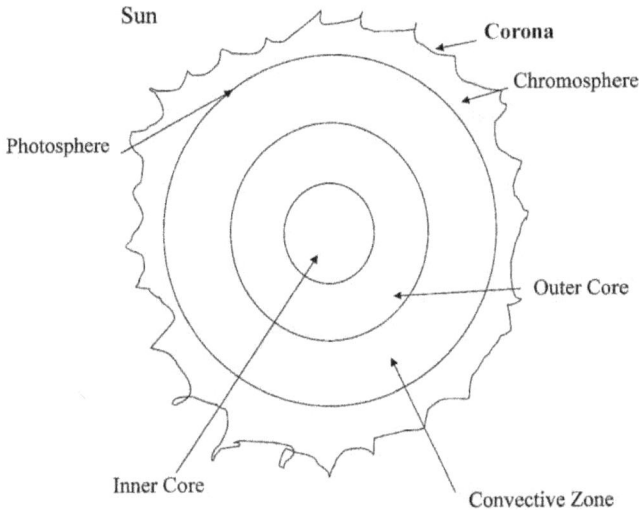

Sun

Corona

Chromosphere

Photosphere

Outer Core

Inner Core

Convective Zone

Diagram 3-1

Scientists divide our sun into inner and outer cores, a convective zone, photosphere, chromosphere, and corona. Generally, as you move outward from the core, temperature decreases. But, for some reason we don't understand, in the outer two layers temperature increases.

The temperature of the chromosphere reaches 18,000 degrees F. The outermost layer, the corona sizzles at up to 1.8 million degrees F. Why is this? About all that most scientists can agree on is that it has something to do with the Sun's incredibly strong magnetic field.

All modern power generation facilities work on the principles investigated by Michel Faraday and put in mathematical form by James Clerk Maxwell. A changing magnetic field produces an electric field. Water or steam turn turbines, which pass powerful magnets through coils of wire to produce the electricity that makes our everyday lives possible. But the opposite is also true. A changing electric field produces a magnetic field. Heated ionized gases rising from the inner regions of the Sun, cooling and falling back down to start the process all over again, in essence turn the Sun into one huge dynamo, a giant magnet.

Take a bar magnet and place it under a piece of paper and sprinkle iron filings on the paper. The filings clump in large looping lines from pole to pole. When one of these lines of magnetic force breaks through the surface of the Sun (illustration 3-1), it inhibits the flow of heat to that point. The dark spots Galileo saw 400 years ago, that someone with the proper equipment can still see today, are not the parting of the Sun's atmosphere to reveal the dark surface beneath. Sunspots are not cold at all. They are just cooler than their surroundings and therefore appear dark. Because lines of magnetic force occur only in closed loops that run from pole to pole sunspots occur in pairs.

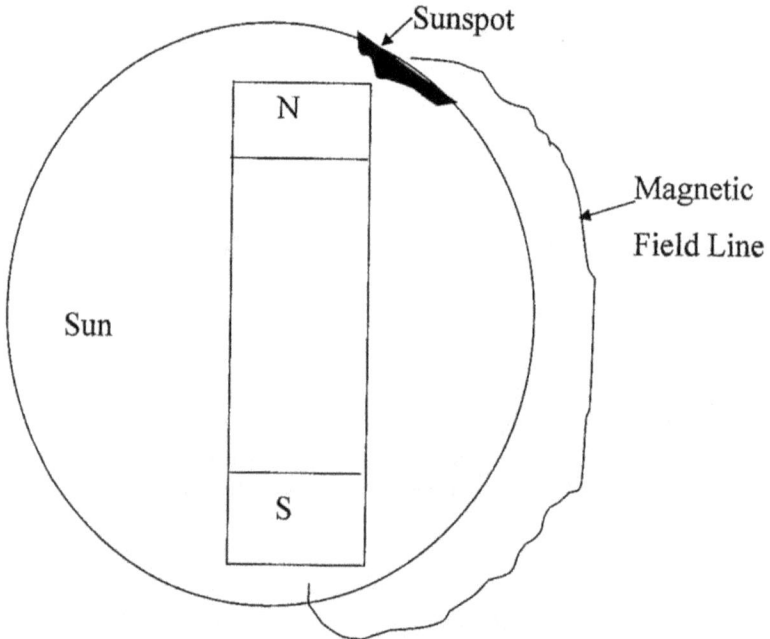

Illustration 3-1

The flow of charged particles in the sun, due to convection, causes it to behave like a giant bar magnet. Where magnetic field lines breach the surface it becomes cooler. The result is a sunspot.

Most of what we know about the Sun and other stars comes from the science of spectroscopy, pioneered by the likes of Gustav Kirkoff, and Robert Bunsen, and Joseph von Fraunhofer, and applied to astronomy by William Huggins in 1862 (see Chapter 5). When the light from a telescope is passed through a prism or diffraction grating, a thin film with millions of tiny, parallel slits cut in it, it produces a spectrum or rainbow of colors. When this spectrum is magnified, all sorts of lines appear (diagram 3-2). These are the same lines produced when, as an example, sodium is burned over a flame, or in later years when gas in a tube is excited by passing a current through it and the resulting spectrum is magnified. In other words, these lines are the fingerprints of elements.

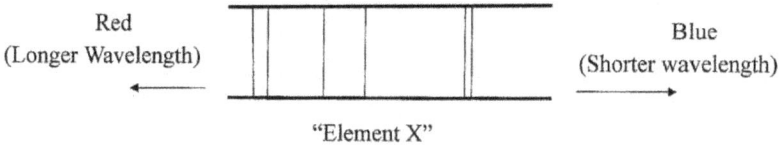

Red
(Longer Wavelength)

Blue
(Shorter wavelength)

"Element X"

Diagram 3-2

Spectrograph of "element X." When passed through a prism or diffraction grating and magnified lines (either dark or bright) appear in an element's spectrum. In stellar or galatic sprectra, these can be used to determine elemental composition and abundance, temperature, magnetic and electric field strength, and (from the spectral shift) whether the source is moving towards (blue-shift) or away (red-shift) from us.

Astronomers can garner a wealth of information from examining a star's spectral lines, their width, shape, and whether what appears to be one line is actually two. To the eye of a trained observer or a properly programmed computer, these characteristics reveal temperature, abundance of the particular element, and magnetic and electric field strength. Astronomers use this information to classify stars into distinct spectral classes: O, B, A, F, G, K, M from hottest to coolest. Generations of astronomy students have memorized this sequence using the saying, "Oh Be A Fine Girl Kiss Me."

O-type stars are much more massive than our Sun, with surface temperatures of around 30,000 degrees Celsius. The Sun is a G-type star, with a surface temperature of about 5,000 degrees Celsius. The Sun is also a second generation star, or one that was formed from the ashes of a previous star. This is indicated by the high metal content in the Sun's spectrum, metal to an astronomer or physicist being anything on the periodic table heavier than helium. At the opposite end of the spectral classification scheme would be the large, relatively cool red star Antares of Taurus.

Astronomers categorize the brightness of a star by its magnitude. The most familiar system of magnitude would probably be earthquake magnitude. A magnitude 3 quake is ten times as powerful as a magnitude 2, and 10^2 or 100 times as powerful as a magnitude 1 quake. Such scales are called logarithmic. Stellar magnitude is also a logarithmic scale, but instead of increasing or decreasing by a factor of ten, each step on the scale changes by a factor of about 2.5 (actually closer to 2.512). This was established by the British astronomer N.R. Pogson.

Unlike earthquake magnitude, the smaller the number the brighter a star. A magnitude 2 star is about 2.5 times brighter than a magnitude 3 star. A magnitude 2 star is 2.5^2 times brighter than a magnitude 4 star, and so on. The brightness of a star as seen from Earth is called its apparent magnitude. The brightness from a distance of ten parsecs (see Appendix C) is called the star's absolute magnitude. A number of other types of magnitude will not be discussed here because in general they are of no interest except to professional astronomers.

When the Danish astronomer Ejnar Hertzsprung and his Princeton counterpart Henry Norris Russell plotted the spectral type of stars on the horizontal axis of a graph, and their luminosities on the vertical axis they found that the majority of stars fell within a strip running from the upper left to the lower right of the graph (diagram 3-3). This is the aforementioned main sequence. The H-R diagram is remarkable in that it presents the whole of stellar evolution in a single glance.

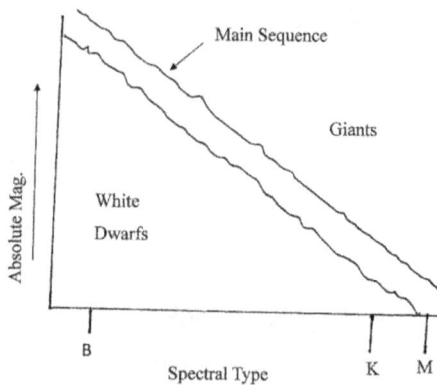

Diagram 3-3

The Hertzsprung-Russel diagram plots a star according to spectral type on the horizontal axis, and magnitude (brightness) on the vertical axis. The majority of stars fall within a strip from upper left-lower right called the "Main Sequence."

A cloud of gas and dust collapses. If the cloud is big enough and if the right temperature and density are reached, fusion begins in the cloud's core, and a star is born. Eventually, as described previously, it enters the main sequence where pressure and the force of gravity balance each other. Where it enters and how long it remains depend largely on the star's mass. As the saying goes, all good things must come to an end. The supply of hydrogen dwindles, and the star leaves the main sequence.

Stars can meet their ultimate fate in one of several ways, again depending on their mass. In low mass stars like the Sun, helium steadily accumulates in the core with hydrogen continuing to burn in a shell around it. The star cools, and its outer layers expand. This is the red giant phase of stellar evolution. When the Sun becomes a red giant some 5 billion years from now, its atmosphere will engulf the inner planets, including Earth. The atmosphere will disappear, the oceans will boil away, and the planet will become a lifeless, irradiated husk.

Eventually, the helium will begin to fuse into carbon. Carbon builds up in the core, followed now by a shell of helium and hydrogen beyond that, like the layers of a cosmic onion (diagram 3-4). Finally the star gives one last gasp, or cough, gently blowing away its outer gaseous layers, which then begin to glow giving birth to a new planetary nebula. The most famous example of a planetary nebula would probably be the Ring Nebula in the constellation of Lyra.

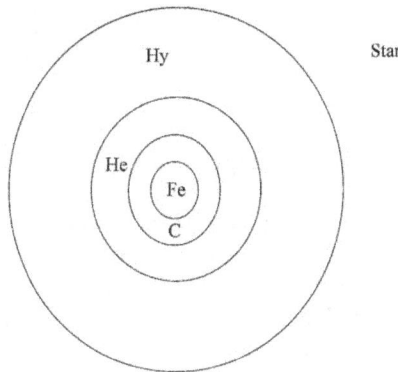

Diagram 3-4

As helium accumulates in the core of a star, it begins to fuse while hydrogen fusion continues in a layer around this. If the star is large enough, the process continues on through carbon and eventually iron; the layers resembling the layers of an onion.

What's left collapses to a ball about the size of the Earth. The result is a very hot (10,000 degrees Celsius), dense object called a white dwarf. If you had asked an astronomer or physicist prior to 1911 whether such a dense object was possible, they would likely have laughed in your face. Not only because the process of stellar fusion/evolution was not understood as it is now, but also because the anatomy of an atom was a mystery.

Scientists knew that the atom consisted of equal amounts of opposite (positive and negative) charges, but prior to 1911 the accepted model for the arrangement of these charges was the so-called plum pudding model of J. J. Thomson. In this model the atom consists of a positive protoplasm dotted with tiny negative electrons (diagram 3-5), the plums in the plum pudding. The plum pudding atom couldn't be squeezed or compacted because there was no empty space.

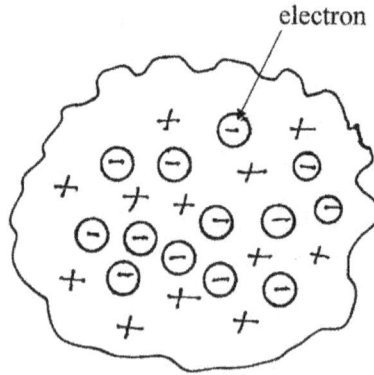

Diagram 3-5

The "Plum pudding" model of the atom as proposed by J.J. Thomson. The atom was seen as sort of a gelatinous mass of positive charge dotted here and there by negatively charged electrons (the plums in the pudding). The plum pudding atom could not be compressed because there was very little empty space.

In 1911, the physicist Ernest Rutherford invented an experiment to test the plum pudding model (illustration 3-2). Using a radioactive source, he bombarded a thin gold foil with alpha particles, or helium nuclei, and watched to see how they were scattered. A screen coated with a substance that would fluoresce when struck by the alpha particles was placed behind the gold foil. If the plum pudding model was correct, then all the alpha particles should be scattered by the foil by roughly equal amounts.

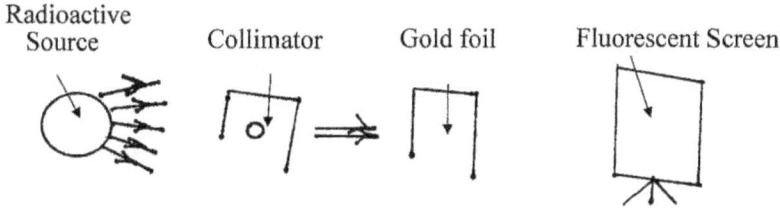

Radioactive Source   Collimator   Gold foil   Fluorescent Screen

Illustration 3-2

Rutherford's experiment for testing the "Plum pudding" atom. Using a radioactive source, and collimator (lead sheet with a small hole in it) he fired alpha particles (helium nuclei) at a sheet of gold foil and watched a screen to see how they were scattered. Some were scattered a little. Some were scattered quite a bit. Some bounced straight back. This would not have happened in the "Plum pudding" atom.

Imagine Rutherford's surprise when some of the particles were scattered hardly at all, and some were scattered a great deal. In fact a graduate student named Hans Geiger informed Rutherford that some of the particles bounced almost straight back. In response, Rutherford penned one of the most famous quotes in scientific history. "It was almost as incredible as if you had fired a 15-inch shell at a piece of tissue paper and it had come back and hit you." The plum pudding model no longer fit the available evidence.

In its place came the nuclear model in which positively charged protons, and later the neutral neutrons, occupied the compact nucleus of the atom, around which orbited the negatively charged electrons. The nuclear atom, in contrast to the plum pudding model, is mostly empty space. When I rest my hand on a table, my hand does not pass through, not because either my hand or the table are solid in the usual connotation of the word, i.e. no space between constituents. Rather solidity is an illusion produced by electrostatic forces. Like charges attract, and opposite charges repel. The electrons on the surface of my hand repel those on the surface of the table, giving the impression of solidity. So the nuclear model of the atom meant that objects like white dwarfs were possible, but no one had yet found one. That would change.

After measuring the parallax of the star 61 Cygni (Appendix C) Friedrich Bessel turned his attention to the bright star Sirius of Canus Major. Indeed he found that Sirius was wobbling, but not due to the Earth's changing position around the Sun. He calculated it would take Sirius fifty years to complete one loop. There could be only one other explanation. Sirius was being influenced by the gravity of an unseen object.

The unseen companion of Sirius was finally seen in 1862 by the astronomer Alvan Graham Clark. The small white speck had been hidden in the blue-white glow of Sirius. In 1915 Walter Adams managed to take a spectrum of Sirius's companion and found that it was almost as hot as Sirius itself, but only 1/6,000 as bright. The companion must therefore be much smaller, only 3/100 the diameter of our Sun. Subsequent analysis of the orbit of the companion suggested that its mass was around twice that of the Sun. The first white dwarf had been discovered.

As bizarre as white dwarfs are, there are even stranger ways for a star to end its life. In stars more massive than four times the mass of our Sun, fusion continues past carbon to oxygen and neon. Main sequence stars with greater than eight solar masses continue the process, building layer upon layer of fusion products ending with iron. Iron is THE dead end for all stars no matter how massive, the point of completely diminished returns. It takes more energy to fuse iron into heavier elements than is released in the process.

The iron core grows until its weight can no longer be supported by the pressure of the fusion reaction, and it collapses. The shockwave generated by the core collapse causes an explosion called a supernova. For periods of days or even weeks, supernovae can outshine the entire galaxies of which they are a part.

In the intense heat and pressure of the supernova, the rest of the elements heavier than iron are born. The explosion spreads these elements throughout space to be used in producing stars like our Sun with planets that are capable of supporting beings who are able to in some small way, at least understand the universe they inhabit. Every Forty-niner who struck it rich in the gold or silver fields of California or Nevada, every wealthy uranium speculator, owes his millions to a star that died billions of years ago. Carl Sagan was not being overly dramatic when he claimed that we are all made of star stuff.

One of two things can happen next. At the very least, the remaining material will continue its collapse to the point where the protons and electrons of the constituent atoms meld. The resulting neutron star is so dense that a teaspoon of its material if brought to Earth would weigh a billion tons. Neutron stars rotate very quickly and have extremely powerful magnetic fields. Beams of radiation stream from the magnetic poles. These can be detected by radio telescopes on Earth. At first these interstellar beacons were called LGM (Little Green Men) because they were mistaken for extra-terrestrial signals. Modern astronomers call these objects pulsars. The regularity of their pulses make them the most accurate clocks in the universe. If the supernova remnant is massive enough, the collapse doesn't stop there. It continues until what is left is a point of almost infinite density, a black hole.

If I throw a rock into the air, it eventually falls back down (illustration 3-3). If I throw it a little harder, it goes a little higher, but gravity eventually wins the struggle, and it falls back down (illustration 3-4). If I could throw the rock hard enough, it would escape the Earth's gravity and travel to infinity (illustration 3-5). The velocity at which this happens is called the escape velocity. For the Earth it is roughly seven miles per second. Remember from Chapter 1 or high school physics that the force of gravity at Earth's surface depends on its mass and inversely on its radius. If the Earth were more massive or the radius smaller, the escape velocity would be greater. Squeeze the Earth into a small enough ball, and its escape velocity would eventually exceed the velocity of light. The Earth would become a black hole. The radius of the Earth at this point would be equal to its Schwartzchild radius. The boundary between the point of no escape (since the speed of light is the ultimate cosmic speed limit) and the rest of the universe is called the event horizon.

Illustration 3-3

If I (or you) toss an object in the air, it travels a height, then falls (under the influence of gravity) back down.

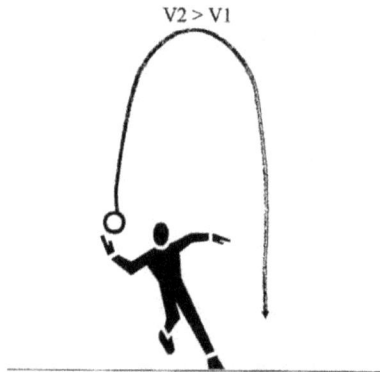

Illustration 3-4

If I (or you) toss the object a little harder, it goes a little higher, but still comes back down.

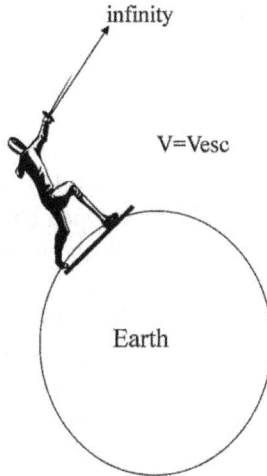

infinity

V=Vesc

Earth

Illustration 3-5

If you were to toss the object with a high enough velocity it would escape the Earth's gravity completely. Scientists call this velocity the "escape velocity." If the Earth were much more dense, (i.e. The same mass squeezed into a much smaller sphere) the escape velocity would eventually become greater than the speed of light and we would have a "black hole."

The incorrect analogy that is most often applied to black holes is that of "cosmic vacuum cleaners." The vacuum cleaner analogy implies just that, a vacuum or negative pressure. This is not the case at all. A black hole will not suck unsuspecting objects into its maw. In the example above, the Moon would go on happily orbiting at the same distance as it did before the Earth was so mercilessly compacted. Life for objects outside the event horizon continues as if nothing had changed.

At this point it might occur to some of you to ask, if black holes are black, how do we know they exist? We must infer the existence of black holes from their effects on their surroundings. When astronomers observe a star orbiting around a massive (at least three solar masses), unseen companion, there is little else the companion could be. Sometimes a black hole will draw matter off a nearby star, which accumulates around the black hole in an accretion disk. As the matter spirals around the black hole, it heats up to the point that it emits x-rays. Increasing evidence suggests that many if not most galaxies, including our own, have super-massive black holes at their centers, and that these may be responsible for the phenomenon called quasars or active galactic nuclei (see the list of books for further reading if you want more details on these fascinating objects). It is even possible that the Milky Way was active at one time.

Most stars form in pairs or even groups known as clusters. Open clusters contain a few tens to a few hundred stars traveling through space together. These tend to be young stars. Open clusters are also called galactic clusters because they tend to congregate in the plane of the Milky Way. Globular clusters on the other hand are dense groups of hundreds, sometimes thousands, of very old stars. For reasons still not fully understood, globular clusters occupy a roughly spherical region around the galaxy's center.

The term galaxy referring to associations other than the Milky Way, did not come into popular use until the 1950s. Early twentieth century photographic plates showed only fuzzy blobs that looked like they might have a spiral structure. The term used to describe these objects was spiral nebulae. Were they or weren't they objects like the Milky Way? Was the Milky Way the entire universe or were there other Milky Ways? This was the crux of the great "island universes" debate.

The debate came to a head at the Astronomical Society meeting held at the University of Chicago in the summer of 1914. The position that the Milky Way was the entire universe, and that the spiral nebulae were just clouds of gas and dust within our galaxy, was defended by Mount Wilson astronomer Harlow Shapley. The opposing opinion, that the spiral nebulae were galaxies in their own right, was taken by Lick Observatory astronomer Heber Curtis. Evidence for either argument was limited, and Shapley and Curtis interpreted some of the same evidence differently. In addition, Shapley claimed that observations made by Adriaan Van Maanen indicated that some of the spiral nebulae were rotating, and that by his calculations if they were indeed galaxies, the stars in the outer reaches of the spirals would have to be moving incredibly quickly, in some cases, even faster than the speed of light, which was obviously impossible (refer to Chapter 4).

The basic argument however could be distilled down to one of size and distance. How big was the Milky Way, and how far away were the spiral nebulae? Here too, as you may have guessed, Curtis and Shapely disagreed. Although both used the period/luminosity relationship of Cepheid variables (see Chapter 4 and Appendix C) to make their estimates, different methods of calibrating the relationship resulted in greatly different numbers for the size of the Milky Way.

Curtis claimed that the Milky Way was just 30,000 light-years across. Shapley argued that the correct number was 300,000. The current accepted value is about 100,000 light-years.

A young Edwin Powell Hubble was invited to attend the meeting. Hubble was a believer that the spiral nebulae were other galaxies. Producing evidence to prove it was the problem. Fortunately for Hubble a new evidence-gathering tool was about to become available.

In 1906 a wealthy Los Angeles businessman named John D. Hooker provided funds for the purchase of a 100-inch glass mirror blank from the Saint Gobain glassworks in Paris. In the era before NASA and the National Science Foundation, that is government funding of big science, most scientific work was made possible by such philanthropic activity. The Mount Wilson 100-inch Hooker telescope did not get off to a very auspicious start. The original blank was cast in three separate pourings, a procedure which allowed for the production of all sorts of bubbles. A second blank cracked during the cooling process. Finally, in 1910, a decision was made to re-examine the first blank.

It was decided that the bubbles were all far enough beneath the surface so as not to present a problem during grinding. Once the polished surface was coated with an optically reflective layer (see Chapter 5), no one would even know the bubbles were there. It took five years to grind the glass blank to the proper shape. A total of one ton of glass was removed in the process. On September 11, 1919, the 100-inch Hooker telescope was cleared for general use.

Dr. Hubble, newly returned from service in the army during World War I, did not get time on the 100-inch right away. There was a pecking order on the mountain and the senior astronomers got first dibs on use of new scope. Hubble spent the majority of his time honing his photographic skills on the 60-inch. Eventually he was allowed time on the Hooker scope.

In a photograph labeled H335H taken by Hubble of the Andromeda nebula with the 100-inch in 1923, he found the evidence he needed. At first Hubble thought he had captured three novae, but upon further inspection, one turned out to be a Cepheid variable. Using Shapley's own calibration of the period/luminosity relationship, Hubble calculated that the Andromeda nebula was 1 million light-years away, well outside the boundaries of the Milky Way. The spiral nebulae were galaxies in their own right. Van Maanen's rotation was a figment of his imagination.

The modern system of galaxy classification is remarkably similar to one that Hubble himself proposed. Galaxies come in four basic types:

spiral, lenticular, irregular, and elliptical. Some of these have subcategories. For instance, spiral can be either normal or barred (see illustration 3-6). The arms of the galaxy may be wound more or less tightly around the central bulge. In order of increasing tightness of the spiral, these are denoted by Sa, Sb, Sc.

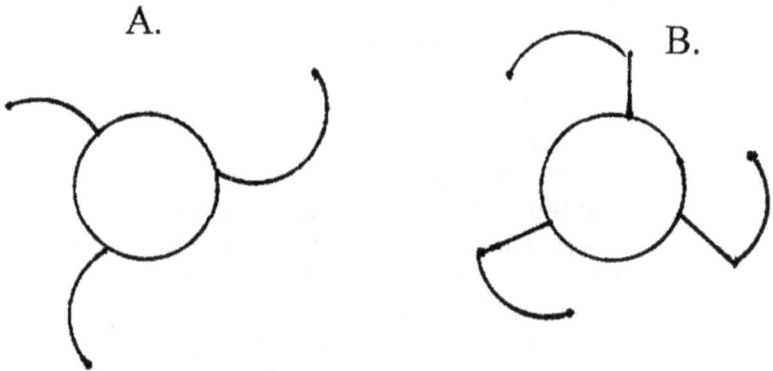

A.

B.

Illustration 3-6

Galaxies are classified as spiral, lenticular, irregular, and elliptical. Within these can be sub-categories as with the spirals which can be normal (A) or Barred (B). In Barred, the arms extend straight out (like the spokes of a bicycle wheel) before curling.

Once thought to be a normal spiral, evidence indicates that the Milky Way may be a barred spiral. Barred spirals are denoted by SB and have arms that emerge from the central bulge like the spokes of a bicycle wheel before curving in the normal manner. Here also the arms may be more or less tightly wound.

Lenticular, or lens-shaped galaxies, have disks with gas and dust similar to spirals, but no spiral arms. Irregular galaxies have little or no structure at all hence their name. Irregulars are also characterized by waves of new star formation. By contrast, ellipticals show little or no evidence of star formation. Ellipticals can actually be spherical to oval shaped, being designated E0-E7 in terms of increasing flatness. Ellipticals may be the result of collisions of galaxies such as will occur between the Andromeda and our Milky Way in several billion years.

Galaxies also congregate in groups. The Milky Way and Andromeda galaxies are part of a cluster known as the Local Group. Clusters in turn form groups or super clusters. In-between these large collections of galaxies are great voids with no galaxies at all. So far no super-duper clusters have been found. As scientists look farther back in space, and therefore farther back in time, with their increasingly powerful telescopes, they continue to find evidence of galaxies, order in an era when there should be none. How is this possible? How did the massive super clusters and great voids form from the relatively smooth Big Bang? Astronomers continue to seek the answers to these questions.

Sources

The Arrow of Time, Peter Coveney and Roger Highfield, Ballantine Books a division of Random House Inc., New York, NY., 1990.

Astronomy For Dummies 2nd edition, Stephen P Maran, Wiley Publishing Inc., Hoboken, NJ., 2005.

The Cosmic Mind Boggling Book, Neil McAleer, Warner Books Inc., New York, NY., 1982.

Edwin Hubble: Mariner of The Nebulae, Gale E. Christiansen, Institute of Physics Publishing, Bristol, UK., 1995.

Man Looks at The Cosmos: The View From Planet Earth, Vincent Cronin, Quill, New York, NY., 1981.

Miss Leavitt's Stars, George Johnson, W.W. Norton & Company Inc., New York, NY., 2005.

Stardust, John Gribbin, Penguin Press, London, UK.,2000.

Stargazer: The Life and Times of The Telescope, Fred Watson, De Capo Press, Cambridge, MA,. 2004.

Sun in a Bottle, Charles Seife, Penguin Group Inc., New York, NY., 2008.

Quasar, Quasar, Burning Bright, Isaac Asimov, Doubleday & Company Inc., Garden City, NY., 1978

CHAPTER 4

THE EXPANDING, ACCELERATING UNIVERSE

From the title you've probably guessed that this chapter is about motion. A good place to start, then, is by asking some seemingly simple questions. Questions such as, how do we measure motion, or how do we know we are moving? "What does he mean?" you ask. "I just know. There's no big mystery." Perhaps an example from life will serve to illustrate what I'm driving at.

I'm willing to bet that almost everyone reading this has had an experience similar to the one that follows. You're driving in your car when you stop for a traffic light. Other cars pull up alongside of you. You're focused on the light, waiting for it to turn green, when out of your peripheral vision you notice something is changing. You try to put the brake pedal through the floor, thinking you are rolling backward, only to realize that it is the car next to you that is moving slowly forward. Now do you see why I asked the question? The answer is, we judge our motion relative to our surroundings.

Let's take a look at another example. After a wild New Year's Eve party you've passed out after a little bit too much to drink. Your friends drag you to the station and put you on a train home. The trains in this town are specially designed so that no engine noise can be heard, and no vibrations can be felt inside the cars. When you wake up you find yourself in a car with just one window facing an identical car on a parallel track.

You desperately need to know if it's safe to get out of the car, because after drinking all that liquor, nature is calling and the restroom in the car you're traveling in is out of order. You look out the one window and notice a passenger in the other car on the parallel track. Luckily he looks your way, and by hand signals, you manage to get the point across that you want to know whether it's safe to get off. The passenger in the other car assures you that it is. You rush over to the door of your car and dive out only to be dashed to bits as both trains go swooshing by.

Did the passenger in the other car purposefully mislead you? No! One of my favorite lines from the Star Wars movie series is from the scene in "Return of the Jedi" where the spirit of Obi-Wan Kenobi tells Luke Skywalker that he will find that, "many of the truths we cling to depend greatly on our own point of view." The same is true of the physics of motion. The passenger in the other car had no more information to go on than you did. Instead of taking time to observe the ground rushing away beneath you as you opened the car door, you dove out, controlled by your bathroom urges and became a victim of classical, or "Galilean" relativity.

Galilean relativity says that on a plane, train, boat, or car moving at constant velocity, a so-called inertial reference frame, there is no physics experiment that you can do to determine whether you are absolutely at rest or moving. Drop a ball in the hold of a boat sailing in smooth seas at constant speed and it will behave exactly as a ball dropped in your laboratory back home. Mount a radar gun on a train moving down the track at 20 miles/hour, and use it to measure the speed of a train coming down a parallel track in the opposite direction at 40 miles/hour and the radar gun will measure the speed of the oncoming train as 20 + 40 = 60 miles/hour. There is no way to measure "absolute" motion, just motion relative to something else.

If that bothers you, join the club. Scientists may not be able to prove things in the sense that mathematicians do (see Chapter 1), but they use math as a tool, and math deals with absolutes. Two plus two is four whether you're flying a plane or riding a bike. Three squared is nine whether you're a college professor or a lowly grad student. The idea that the best anyone can say is, "I'm moving relative to those trees over there, which may themselves be moving relative to something else ad infinitum," was, and still is, disturbing on a deep level.

Another enduring mystery of twentieth century physics concerned the nature of light (see Appendix D). From the time of Newton, light was known to exhibit some properties of waves. Every other wave needed a physical medium to propagate: water waves used water, sound waves used air, earthquake waves propagated through the crust and mantle of the Earth. What medium did light use? Scientists resurrected an old friend.

The aether, once used as an explanation for the composition and motion of the stars and planets (see Chapter 1), might hold the key to the propagation of light. This luminiferous aether would have to have some very strange properties. It must pervade all of space. It must be rigid enough to provide a medium for the propagation of light, yet insubstantial enough so as to not provide any resistance to the motion of the planets around the Sun, or the Moon around Earth. It might also finally provide a true rest frame by which absolute motion or absolute rest could be measured. Motion relative to the aether would be "absolute" motion. How could scientists tell?

If the aether was the means by which light traveled, then it should be possible to detect the Earth's motion through the aether by looking for its effect on the speed of light. In 1887, Albert Michelson and Edward Morley designed an experiment to look for this effect using an interferometer. A light beam was split in two by a half-silvered mirror, or beam splitter, sending the resulting beams down two perpendicular paths of equal length. At the end of each of these paths was a mirror (see diagram 4-1). The beams were reflected back and recombined. As the Earth moves through the aether the two beams should have been affected by their passage through the aether, like a swimmer swimming against the current as opposed to swimming perpendicular to the current. Like the previous example of the train, the speed of light measured in the direction of the Earth's motion should be the speed of the Earth around the Sun plus the speed of light measured by an observer at rest with respect to the aether. When the beams recombined they should have interfered to produce a series of light and dark lines on a screen (see the discussion of interference in Chapter 6).

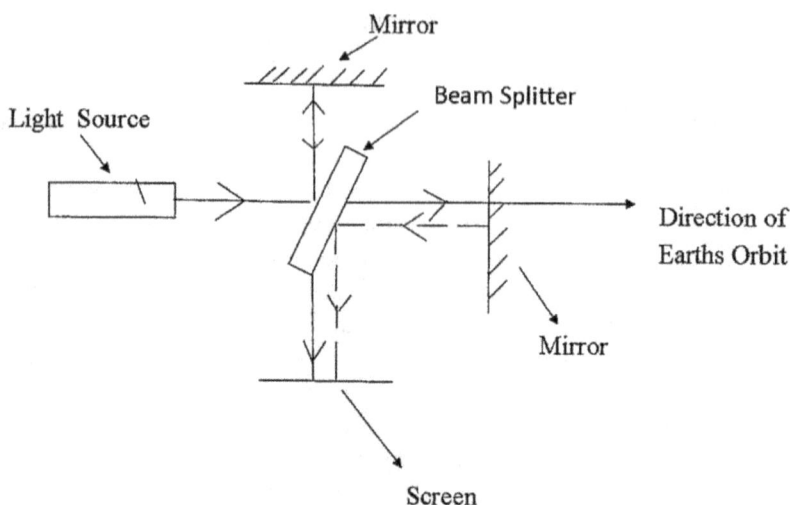

Diagram 4-1

The Michelson-Morley experiment for detecting the presence of the "luminiferous aether." Light from a source is split and travels along two paths of equal length perpendicular to each other. The difference in the speed of light along the paths, one in the direction of the Earth's motion, the other at right angles to it, should cause the beams to interfere when recombined. Michelson and Morely found no such interference despite every precaution.

Michelson and Morley went to great extremes to ensure the accuracy of their experiment. To eliminate the smallest of vibrations, they floated the entire apparatus in a pool of mercury. Because it was difficult to make each arm of the interferometer exactly the same length, they decided to rotate the apparatus and look for changing interference patterns rather than just the presence of the patterns themselves. But as careful as they were, Michelson and Morely could find no evidence of the aether. Light, it seemed, did not play by the rules of Galilean relativity.

What made light so special? Different scientists naturally had different thoughts on the matter. One name stands out, however: Albert Einstein. According to Einstein himself, he first became interested in physics at an early age when a relative gave him a small compass. Young Albert was fascinated that an unseen force could cause the needle of the device to move. Einstein grew up to become the undisputed king of the "gedankan" experiment, literally thought experiment. Experiments that take place completely in one's mind, like our example of the drunken train passenger, thought experiments are no less scientifically valid than experiments that take place in a lab. Unfortunately Albert's early teachers didn't appreciate his abilities, branding him as nothing more than a hopeless daydreamer.

Sometime in his teens, Einstein began to think about the nature of light. What if it were possible to travel alongside a light wave? In the late 1800s, James Clerk Maxwell introduced his equations of electromagnetism. These predicted that light was an electromagnetic wave, composed of an electric and magnetic part. If you could speed alongside a light wave, would you see the individual electric and magnetic parts? Einstein didn't think light would be light if that were possible.

Other strange things would happen if light followed the rules of Galilean relativity. Effects would precede causes. Say Mr. Golfmore wants to play a few rounds one sunny morning. Since he lives just a few blocks from the country club, he decides to just drive the golf cart. As Mr. Golfmore nears the intersection, he sees young Johnny Lightyear approaching from his left (see Diagram 4-2) riding his skateboard and listening to his iPod, oblivious to his surroundings. Mr. Golfmore swerves just in time to avoid knocking Johnny out of his earphones.

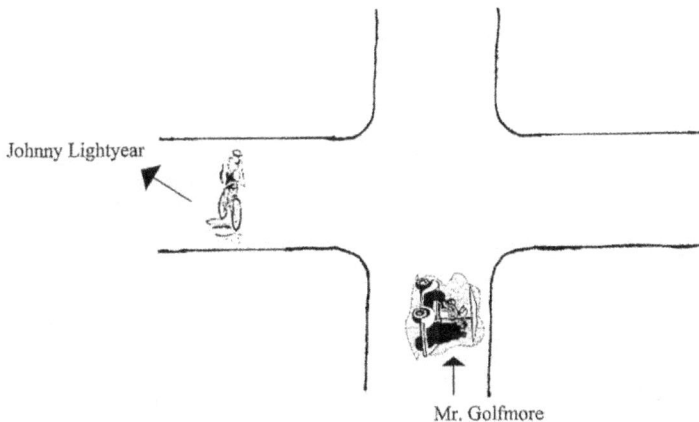

Diagram 4-2
Mr. Golfmore decides to drive his cart to the course to play a few rounds. He sees Johnny Lightyear approaching from the left. Mr. Golfmore swerves just in time to avoid Johnny, (or so he thinks).

A simple enough scenario. What's the big deal? Taking the advice of Obi-Wan Kenobi, let's look at it from a different point of view. Mrs. Bright happens to be walking her dog, Warpspeed, and is standing on the corner facing Mr. Golfmore when the incident occurs (see diagram 4-3). If Mrs. Bright perceived the speed of light reflecting of Mr. Golfmore's golf cart to be the speed of light reflecting off a stationary cart plus the speed of the cart, then she would see him swerve for apparently no reason. The light reflecting off the golf cart would reach her before the light reflecting off of Johnny.

Johnny Lightyear

Ms. Bright

Mr. Golfmore

Diagram 4-3

Mr. Golfmore decides to drive his cart to the course to play a few rounds. He sees Johnny approaching from the left. Mr. Golfmore swerves just in time to avoid hitting Johnny, (or so he thinks). Ms. Bright sees Mr. Golfmore swerve for no reason because the light reflecting from the golf cart reaches her before that from Johnny. Effect precedes cause.

So if the speed of light remains the same no matter the speed of the source or of the observer, what gives? One more thought experiment and a little math (forgive me) shows that the fault lies not in our ourselves, but in our measuring devices. We start with a railroad boxcar with one side removed so that we can see the contents. Inside is a device, a laser perhaps, that fires a beam of light at a mirror on the ceiling of the boxcar (see illustration 4-1). If we know the distance from the ceiling to the floor, say d, then it is easy to calculate the time it takes the beam to travel from the floor to the ceiling and back. It's just the distance, 2d, divided by the speed of light, denoted in most physics textbooks by the letter "c."

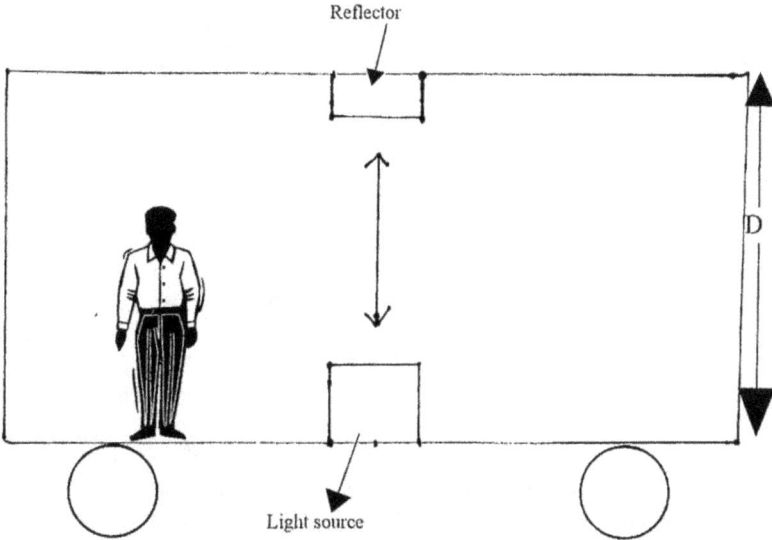

Illustration 4-1

Part 1 of Einstein's special relativity thought experiment. A beam of light is fired from a device on the floor of a stationary box car. If the speed of light is C and the height of the boxcar is D, then the time the beam takes to bounce off a mirror at the top of the car and return to the source is just 2d/c.

Now connect the boxcar to an engine pulling it with a speed v. As the boxcar passes an outside observer, the laser is triggered. To an observer inside the boxcar, the light travels the same path as before. To the observer standing alongside the track, the light wave travels a diagonal path (see illustration 4-2). With nothing more than a little high school trigonometry (anyone interested in the details is encouraged to look them up in any freshman college physics text), some startling results reveal themselves. The speed of light stays the same, but the rulers and clocks of different observers are different.

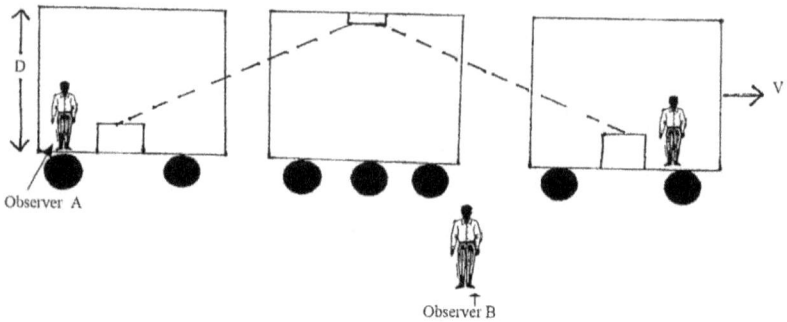

Illustration 4-2

Second part of Einstein's special relativity thought experiment. The
boxcar is moving with speed V. To "observer A" the light travels the
same path as before, taking time 2d/c. To "observer B" standing
outside as the boxcar goes by, the light appears to travel a diagonal
path, but the only thing that has changed is the point of view or
"reference frame" of the observer. Since the speed of light doesn't
change, Einstein concluded that the observer's measurement of time
and distance must. The idea of "time dilation" and "length
contraction" were born.

    As measured by the observer standing alongside the track as the
boxcar goes by, a yardstick in the boxcar does not measure a yard, but
rather something less than a yard. How much less is determined by a
factor denoted by the Greek letter Gamma, γ, which depends on the ratio
of the speed of the boxcar squared, to the speed of light squared. The
effect by which the lengths of moving objects are shortened is called
length contraction. Of course, the observer in the boxcar claims he is the
one at rest and therefore it is the yardstick of the observer alongside the
track, which is askew. This leads to some interesting paradoxes, which I
leave it to the reader to discover (see the list of works for further reading
at the end of this book).

    Not only are lengths of moving bodies contracted, but a clock on
the moving boxcar ticks more slowly relative to a clock alongside the
track by the same factor of Gamma. This effect is called time dilation.
What's more you cannot have one effect without the other. You cannot
experience length contraction without associated time dilation. Space and
time are connected. As a result of Special Relativity it was impossible
thereafter to think of three dimensions of space and a separate dimension
of time. Rather, one had to now think of four-dimensional space-time.

Einstein was still a lowly patent clerk when he published his Special Theory of Relativity. It was special, not in the sense of being super keen, or necessarily better than any other theory, but rather because it covered the special case of uniformly moving systems, or inertial reference frames. This is the same theory that gives us the equation most often associated with Einstein, $E=mc^2$. Although this is the equation behind the atomic bomb, and the one that tells us how stars get their energy, Einstein did not win the Nobel Prize for Special Relativity.

As fantastic as the predictions of Special Relativity are, every time they have been tested they have been confirmed. Like most theories, though, it not only answered some questions, but posed new ones. Having established the speed of light as the ultimate speed limit, the Newtonian theory of gravity was badly in need of an overhaul. According to Newton, gravity was an innate property of matter that made itself felt instantaneously at potentially infinite distances. According to Newton, if the Sun were to suddenly disappear, the Earth and the other planets would immediately go flying off into space. But what about that cosmic speed limit? Einstein showed that no information could travel faster than the speed of light. Neither the Earth, nor anyone on it, should know for about eight minutes, the travel time of light from the Sun, that the Sun had disappeared.

Einstein went back to what he was good at, the thought experiment. What properties did gravity possess? It was an attractive force, always pulling objects together. The more massive the object, the greater its gravity. But was it actually a force? Imagine standing inside a closed container (an elevator car is the most commonly used example). You drop a ball and watch as it falls to your feet. Exactly what you would expect if the container were sitting at the surface of the Earth. It's also exactly what you would expect if the container were in outer space being pushed along with rocket motors with an acceleration equal to one g, the acceleration due to gravity at the surface of the Earth. A physicist would say that it's possible to mimic the effect of gravity by a suitable change of reference frame (a container being accelerated by rocket motors).

Likewise, it's possible to negate gravity by a change of reference frame. Imagine yourself in the same container suspended from a crane at the top of a really tall skyscraper. At some point the container is released. It begins to fall, so we know gravity is working, but inside the container you feel weightless. If you let go of the ball it would just hover, for as long as the container is falling, that is. The "force" of gravity has been negated by the choice of a non-inertial reference frame. This should not be possible with a true force of nature. Einstein felt very strongly that the true forces and laws of nature should be the same whatever your reference frame, i.e., state of motion. But what was gravity, then, if not a force of nature?

The mathematics of Einstein's second relativity theory is much more complicated than the math behind his first. In fact it's generally reserved for graduate school, but a few more thought experiments should serve to illustrate the basic ideas. Now suppose that from the inside of the falling container you drop two balls from your outstretched arms. For a while they seem to maintain the separation between them, but gradually the distance between them shortens as if some mysterious force, like gravity, is pulling them together (see illustration 4-3). Actually, gravity is working to pull both of the balls and the container towards the center of the Earth, but from the inside of the container you have no way of knowing this. What else might produce such an effect? How about a curved surface?

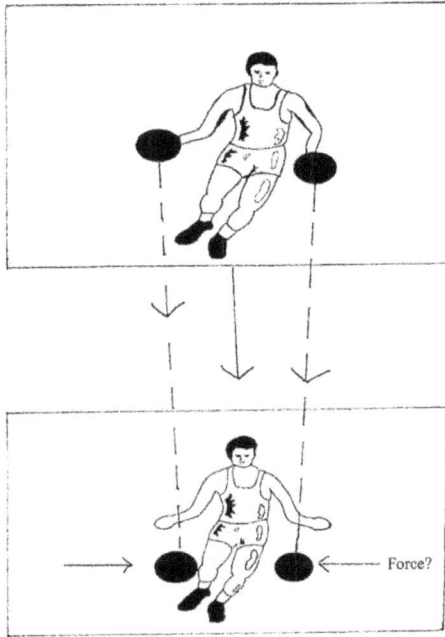

Illustration 4-3

A container (perhaps a shipping container), is dropped from a considerable height. Inside the container our subject releases a ball from each hand. In free fall he/she feels weightless. For a time the balls maintain their distance but container, subject, and balls fall toward the Earth's center. The balls inch closer together as though some "force" is pushing them together. There is no such force. Likewise, Einstein concluded there is no such "force" as gravity.

If you and a neighbor several blocks apart start walking due north, you will appear at first to maintain your distance. Eventually, as you get closer to the pole, the distance between you will shorten until you finally meet at the pole. Was some strange force drawing you together? No. You end up in the same spot because you were walking on a curved surface. What if gravity were nothing more than an affect of curved space-time? Einstein didn't publish his General Theory of Relativity, until 1917, but when he did it was a revolution in our understanding of the universe.

Place a bowling ball on a trampoline. The ball distorts the fabric around it. Now take a marble, place it near the bowling ball and release it. It moves towards the center of the depression made by the bowling ball, just as an object dropped at the surface of the Earth moves towards the Earth's center (see illustration 4-4). Take another marble, or the same one, and this time give it push as you release it. The marble travels a curved path around the bowling ball. Give the marble just the right push and it will make a complete circle, or one orbit, around the bowling ball (see illustration 4-5)

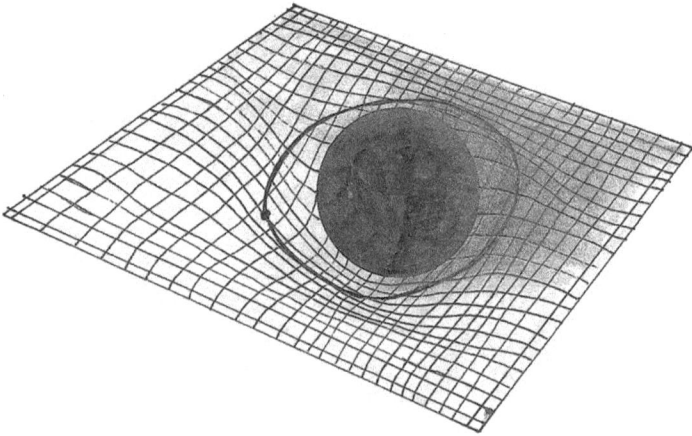

Illustration 4-4

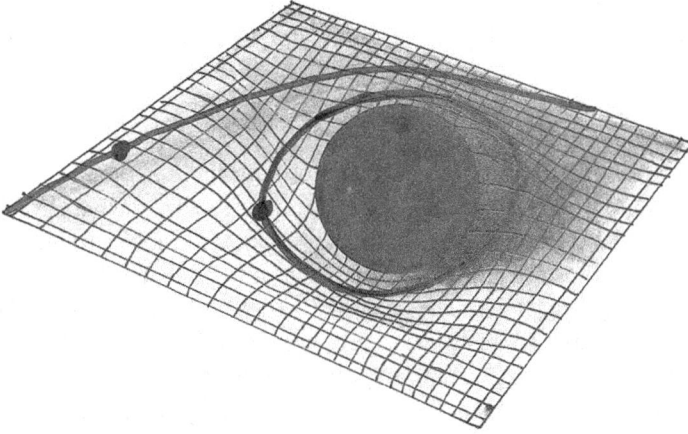

Illustration 4-5

In Einstein's general theory, general because it deals with all reference frames, not just inertial ones, acceleration can mimic gravity, and what we experience as gravity is just the result of matter warping space-time. General relativity predicts strange things, such as the bending of light by a gravitational field, and clocks that run more slowly at sea level than on the top of mountains. Again, every time we've tested the predictions of the theory, they are confirmed. Lest you think that the consequences of the special and general theories have no real-world applications, if you've ever used a GPS (Global Positioning System) locating device, you have taken into account the implications of relativity.

Perhaps the most startling and spectacular consequence of general relativity revealed itself when Einstein attempted to apply the theory to the universe as a whole. His equation would not permit a static solution. In other words, the equation was telling Einstein that the universe could either be expanding or contracting, but it could not remain the same. Everyone, including Einstein, knew that this was preposterous, the same way that a moving Earth was preposterous to pre-Copernican astronomers. So what did the world's greatest scientist do when faced with such a staggering contradiction between what his theory was telling him and common sense? He fudged! Einstein added a term to his equation, the "cosmological constant," to counteract the expansion or contraction of the universe.

The man who would once and for all show the world the error of Einstein's way, Edwin Powell Hubble, was born in 1889 in Marshfield,

Missouri. After obtaining his BS at the University of Chicago, Hubble won a Rhodes Scholarship to study law at Queen's College, Oxford. However, the lure of astronomy was too much for him. He returned to Chicago to get his PhD, and after World War I joined the staff of Mount Wilson Observatory in 1919.

Mount Wilson, near Pasadena, California, had the largest aperture telescope at the time, the 100-inch reflector (see Chapter 3). This was just what the doctor, in this case Doctor Hubble, ordered for his studies of the "spiral nebulae." One of the major debates at the time was whether these objects, appearing as just fuzzy blobs in previous photos, were blobs of gas and dust within our own Milky Way, or galaxies in their own right, although the term galaxy didn't come into common usage until the 1950s. Hubble settled the debate by photographing individual stars within these nebulae, but not just any kind of star, rather very special stars called Cepheid variables.

It is a sad fact that throughout most of the history of science, women have been regarded as having little in the way of original material to contribute, or of being too emotional to do objective research. Most universities refused to grant women advanced degrees in the sciences until well into the twentieth century. Most women were confined to the equivalent of intellectual sweatshops. These women were known as computers. In an era before the electronic version, the human versions were charged with the same mind-numbing tasks: endless repetitive calculation, graphing of data, and anything else deemed too menial for a Ph.D. to bother with.

Fortunately this did not deter all women from pursuing science, or a few from making very significant contributions. One of these was Henrietta Leavitt. As a computer at Harvard Observatory in 1880, she spent her time cataloguing star positions, colors, and brightness for a salary of twenty-five cents an hour and a month's vacation. Later she moved to Boston's Cambridge Observatory where she started cataloguing variable stars. As the name implies, variable stars vary regularly in brightness.

By 1904, Ms. Leavitt had discovered several special variables in both the Small and Large Magellanic Clouds, satellite galaxies to the Milky Way. The Cepheid variables got their name from the constellation Cepheus where they were first discovered. They have a characteristic pulsation that lasts from a few days to a few weeks. Because all of the variables that Henrietta found were in the Small Magellanic Cloud, she could safely assume that they were all at approximately the same distance from Earth. She also noticed that a Cepheid's period of variation seemed to be related to its brightness, the brighter variables having a longer period.

So what? So, if Leavitt was right about the period-luminosity relationship of the Cepheids, then all anyone would have to do to measure the distance to a cluster or galaxy containing a Cepheid would be to observe the period and from that determine the absolute brightness of the star and compare that to its observed brightness (see Appendix C). A light bulb of a certain brightness at one distance will have one-forth that brightness at a distance double that of the first. Triple the distance and the bulb will appear one-ninth as bright. A Cepheid with the same period as another that appears one-forth as bright is twice as far away from Earth. The problem then is to measure the distance to one Cepheid.

In October of 1923, Hubble found Cepheids in the Andromeda nebula. Based on the Cepheid's period, Hubble calculated the distance to Andromeda to be around one million light-years. In one fell swoop, the size of the universe tripled. As amazing as that discovery was, an even greater one was still to come.

In 1912, Vesto Slipher, director of Lowell Observatory (see Appendix A), passed the light coming from the Andromeda nebula through a spectroscope. A spectroscope separates light from an object into a rainbow of colors, a spectrum, and then magnifies it to reveal light (emission) or dark (absorption) lines. These are the fingerprints of elements. Each element has its own unique pattern of lines, and from the width, brightness, and other characteristics of these, astronomers can tell many things about light source. In particular, sometimes the lines are shifted from their normal positions (see diagrams 4-4 and 4-5). A shift towards the red, or longer wavelength end of the spectrum, indicates a source that is moving away from the observer. Likewise a blue-shift indicates a source that is approaching the observer. This is analogous to the effect discovered in 1842 by Christian Doppler in relation to sound waves, and which is experienced by every NASCAR racing fan as he sits in the stands watching the cars go by. Slipher found that the light from the Andromeda nebula was blue-shifted.

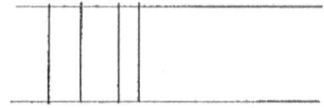

Element X   Element X

Unshifted   Red-shifted

Diagram 4-4

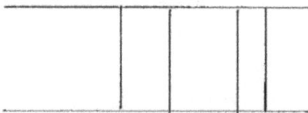

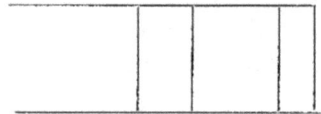

Element X   Element X

unshifted   Blue-shifted

Diagram 4-5

Like the spectra of stars, the spectra of galaxies can be red or blue shifted, indicating motion away or towards Earth. When Vesto Slipher passed light from 25 "spiral nebulae" or galaxies through a spectroscope, he found the majority showed red-shifted spectra. The universe seemed to be expanding.

By 1917, Slipher had obtained spectra from 25 nebulae. What he found was nothing short of spectacular. The majority of the spectra were red-shifted. It seemed the universe was indeed expanding. Still, not all astronomers were convinced. However, during the 1920s, Hubble's partner, Milton Humason, who started as a mule driver delivering supplies for Mount Wilson Observatory's construction, had confirmed and extended Slipher's red-shift measurements. By combining this data with his distance estimates based on Cepheid measurements, Hubble found that a nebula's (galaxy's) recessional velocity was proportional to its distance. The farther away a galaxy was, the faster it was moving away from us. What's more, when the data was plotted on a graph with distance on one axis and velocity on the other, the closest fit was a straight line. A galaxy's speed appeared to be equal to some constant, forever immortalized as Hubble's Constant and denoted by H, or sometimes H_0, times the galaxy's distance from us.

Einstein met Hubble in 1930 when the physicist visited Mount Wilson. When he was confronted with Hubble's data, he had to admit that adding the cosmological constant to his equation had been a mistake. He would later call it the greatest mistake of his career.

The implications of an expanding universe on cosmology were incredible. In grade school, whenever the teacher would show a motion picture, we would always beg him/her to run it in reverse. The sight of people walking backwards, cars driving backwards, or raindrops rising up from the ground would always elicit a giggle. What would happen if we could run the movie of the universe backwards? If most of the galaxies are moving away from each other now, if we run time backwards, they must get closer together, until at some point all the matter and energy in the universe must occupy a single point. The universe must have had a beginning in a tremendous explosion, the Big Bang.

"Big Bang" was actually coined as a derogatory term by astrophysicist Fred Hoyle for what he considered the absurd idea of a universe with a beginning. Indeed, due to troubles calibrating the Cepheid yardstick (see Appendix C), the first estimates placed the age of the universe as younger than most of the stars occupying it. Hoyle and some colleagues preferred the Steady State theory. According to the steady state, matter was continuously being created in the space between the galaxies. Although it has not stood the test of time to become the standard accepted cosmological model, the steady state was a good theory in the sense that its predictions were disprovable by observation. According to the steady state the appearance of the universe should not have changed much over time.

Working on the problem of nucleosynthesis, or creation of elements in the Big Bang, George Gamow and Ralph Alpher predicted that we should still be able to see the radiation left over from the moment of creation. The very early universe was an immensely hot soup of subatomic particles and electromagnetic radiation, or photons. This early universe was opaque. We will never be able to see it, because the photons couldn't travel far without being absorbed by electrons in the plasma. Eventually the universe cooled down enough so that electrons could attach themselves to protons to form atoms. At this point radiation could flow freely. Gamow and Alpher calculated that the expansion of the universe should have cooled the radiation to a temperature of about three Kelvin, or –270 degrees Celsius. This would place the radiation in the microwave region of the electromagnetic spectrum.

Showing that even scientists have a sense of humor, Gamow convinced Hans Bethe, who had contributed nothing to the research to nonetheless sign his name to the paper. The paper has gone down in history as the "alpha, beta, gamma" paper, after the first three letters in the Greek alphabet corresponding to the initials of the authors' names. It was published on April Fools' Day, 1948. The only joke, however, was on those who did not take the paper seriously.

In 1962, astronomers Arno Penzias and Robert Wilson began using a large horn-shaped antenna, originally constructed for long distance satellite communication, for radio astronomy. Their observations were plagued by a persistent source of noise. It seemed to have the same intensity no matter in what direction the antenna was pointed. Penzias and Wilson even went to the extreme of cleaning the pigeon droppings out of the antenna. Nothing worked.

Slowly the pair came to the conclusion that the problem was not in the instrumentation, but might perhaps be a true cosmic signal. Perplexed, they contacted Robert Dicke at Princeton. Dicke had been attempting to detect Gamow and Alpher's radiation unsuccessfully for years. When he heard Penzias and Wilson describe their strange signal, he knew immediately what they had discovered.

Discovery of the cosmic microwave background radiation would win Penzias and Wilson the Nobel Prize in physics, and was the virtual death knell for the steady state theory. About 13 billion years ago, every bit of matter and energy we see around us started at one amazingly hot, infinitely dense point called a singularity. Our laws of physics fail us at these temperatures and densities (see the discussion of black holes in Chapter 3), so the conditions in the first few seconds after the Big Bang are a mystery to us. Where did the Big Bang take place? Everywhere and nowhere. The Big Bang was not a conventional explosion taking place in space at a certain time, but rather space and time were created in the Big Bang.

As mysterious as the universe's beginning is, not even considering what happened before the Big Bang, its end is equally difficult to predict. Throw a rock into the air and gravity immediately begins pulling on it. The rock progressively slows down until at some point it stops and falls back to Earth. Throw it hard enough and it escapes the gravity of Earth and eventually comes to rest at infinity. Throw it harder still, and the rock will continue to travel away from Earth forever. Likewise, the universe could have one of three fates depending on the amount of matter and energy within it.

Shortly after the moment of creation, gravity started working to slow the expansion down. If there was enough matter in the universe, the expansion would eventually stop and reverse, perhaps ending in a Big Crunch. Slightly less mass and the universe will expand to a point and stop. Less mass still, and the universe will continue to expand forever, the stars burning out, space eventually filled with a uniformly cold soup of dust and radiation. Astronomers have named these scenarios the closed, flat, and open universes respectively.

Which universe do we live in? How can we tell? Theoretically, very simply by counting up all the mass in the universe and comparing it to the critical mass needed to stop the expansion. The problem is, the matter we see may not be all the matter there is. As early as 1925, Knut Lundmark attempted to weigh the Milky Way and concluded that the visible part was a factor of 100 less than that needed to explain the motion of the galaxy's parts.

About a decade later, Fritz Zwicky observed the motion of galaxy clusters and determined they would fly apart if not for the gravity of some unseen mass. Finally, in the 1970s, Vera Rubin and Kent Ford studied the motion of stars at varying distances from their galactic centers. If the majority of a galaxy's mass were in the form of visible matter and therefore concentrated in the central bulge, then stars farther from the center should orbit the center more slowly than stars closer to the center, just as planets farther from the Sun orbit more slowly than those closer to it. Instead, Rubin and Ford found that the velocities of stars around the center seemed to be fairly constant from the center out to the edge of the galaxy. The universe seemed to be permeated by some form of matter detectable only in its gravitational effect on the matter we could see, hence this matter was dark in the ordinary sense.

What is dark matter? Is it the familiar particles that make stars and astronomers and planets? Or is it some form of exotic matter such as that predicted by certain cutting edge physical theories? The debate continues, but if there is enough of it, it could have a profound impact on the fate of the universe. It turns out, however, that dark matter was only part of the story.

In the quest to push the boundaries of the universe ever farther, astronomers search the skies for objects called type 1a supernovae. These are very special exploding stars (see Appendix C). The prevailing theory is that type 1a supernovae occur in binary systems where one member is a white dwarf. The gravity of the white dwarf draws matter from its companion star. This matter builds up until the white dwarf can no longer support the increased weight, and it collapses. The collapsing mass causes a renewed burst of fusion and ultimately an explosion. Because all type 1a supernovae are produced by a similar process, they all achieve roughly the same peak brightness. This makes them ideal as standard candles to measure distance (see Appendix C).

Whatever category, open, flat or closed, our universe falls into, astrophysicists assumed that the expansion would be slowing down. Based on this they had certain expectations for the brightness of type 1a supernovae at certain distances. Imagine then the surprise when two groups of researchers found that the supernovae they observed were dimmer than expected. There seemed to be only one conclusion to draw. Not only was the universe's expansion NOT slowing down, it was in fact accelerating.

Einstein's greatest mistake of his career may not have been a mistake at all. The difference is that instead of being needed as a term to counteract the expansion of the universe, the cosmological constant may be needed to explain the acceleration. Astronomers coined the term dark energy for whatever was causing the acceleration. Dark, in this case denotes not only its invisibility, but also our ignorance of its form.

Best current estimates, taking into account visible matter, dark matter, and dark energy, are that the universe is of the flat variety. But whatever the ultimate fate of the universe it seems that the one thing that is certain is that not much is certain. The science of cosmology was born out of change. As we learn more about the universe, our ideas regarding its evolution and ultimate fate will continue to change as well.

Sources

Dark Cosmos: In Search of Our Universe's Missing Mass and Energy, Dan Hooper, Smithsonian Books in association with Harper Collins, New York, NY., 2006.

Discovery of Cosmic Fractals, Yurij Baryshev and Pekka Teerikorpi, World Scientific Publishing Co., Singapore, 2002.

In Search of The Big Bang, John Gribbin, Bantam Books Inc., New York, NY., 1986.

Edwin Hubble: Mariner of The Nebulae, Gale E. Christianson, Institute of Physics Publishing, London, UK, 1995.

Miss Leavitt's Stars, George Johnson, Atlas Books, W.W. Norton & Company Inc., New York, NY., 2005.

Simply Einstein: Relativity Demystified, Richard Wolfson, W.W. Norton & Company Inc., New York, NY., 2003.

The Whole Shebang: A State of The Universe Report, Timothy Ferris, Simon & Schuster, New York, NY., 1997.

✸ ◆ ✸

CHAPTER 5

A BRIEF HISTORY OF THE TELESCOPE

Most of us know a telescope when we see one. Some of you reading this have looked through one. Some of you may even own one of your own, but how often do you stop to think about the humble beginnings of the telescope, and how it has become what it is today? First things first, there are two basic types of telescope: refractors and reflectors. Refracting telescopes use lenses, or shaped pieces of glass, to focus and magnify the light that enters them (diagram 5-1). Reflectors, as the name implies, use mirrors to do the same thing (diagram 5-2). So which came first, you may ask, the refractor or the reflector? The answer is the refractor. The reason was that until the invention of modern methods of vacuum deposition of aluminum and polishing of mirror surfaces, it was much easier to cut and grind a piece of glass to the proper shape to make a lens than it was to polish a metal surface to the proper mirror smoothness.

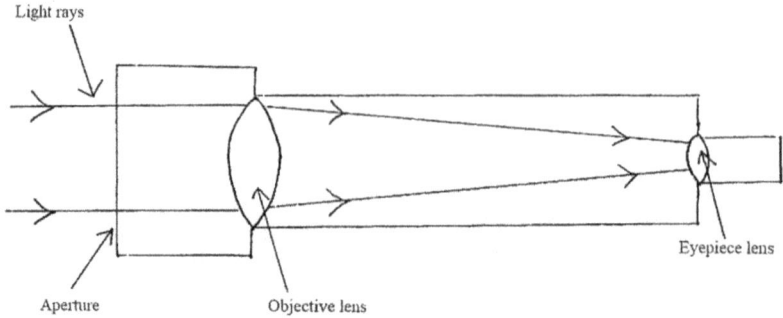

Light rays

Aperture

Objective lens

Eyepiece lens

Diagram 5-1

In a "refracting" telescope, light enters an opening called the "aperture", is bent by an "objective" lens before being magnified by the "eyepiece" lens and entering the eye of the observer. Because it was easier to grind glass than to shape and polish a metal mirror this type of telescope was the first to be constructed.

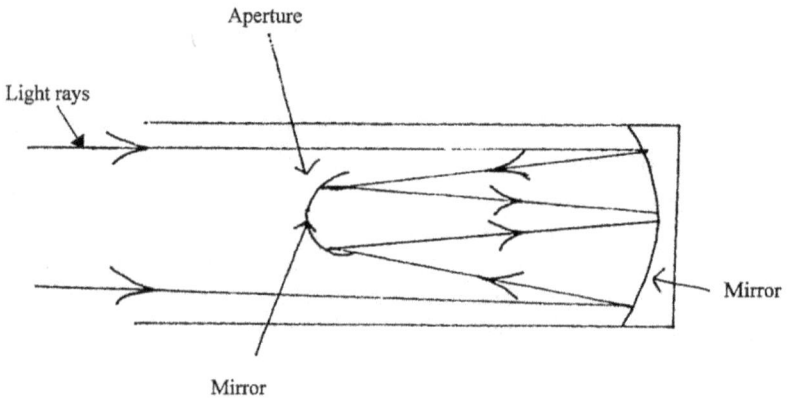

Aperture

Light rays

Mirror

Mirror

Diagrams 5-2

In a "reflecting" telescope, light enters through the aperture, but instead of being focused by a lens, it is focused by bouncing off the curved surface of a mirror. The first mirrors of this type were made of all metal, but because the surface was so hard to polish, images were poor.

Who invented the telescope? We don't know. Many ancient cultures, the Egyptians for one, possessed the ability to cast glass. It was also known that certain lenses could be used to correct for certain vision problems. Convex lenses (illustration 5-1) could be used to help people who suffer from farsightedness. Concave lenses (illustration 5-2) could be used to help people with nearsightedness focus on objects that were far away. However, the name of the person who found that a proper combination of the two could be used to magnify objects many miles away has been lost to antiquity.

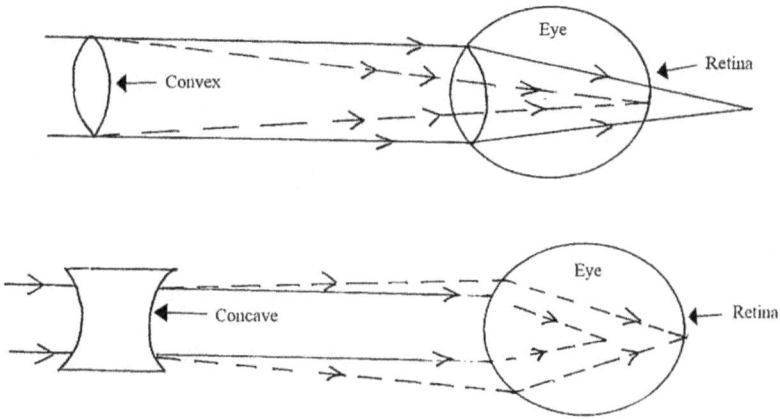

Illustrations 5-1 and 5-2

It has been known for a long time that convex lenses can help people with farsightedness. Likewise, concave lenses can help people with nearsightedness. The name of the person who discovered that a suitable combination of the two types could produce a telescope for magnifying distant objects is lost forever in the mists of history. It was NOT Galileo.

The first record of anyone applying for a patent on the telescope-making process was that applied for by the Dutchman, Hans Lipperhey. On October 2, 1609, Lipperhey asked the States General in The Hague to grant him a thirty-year patent. Unfortunately for Lipperhey, the telescope was an idea whose time had come with a vengeance. Two weeks later another man, Jacob Andriaenszoon, arrived at The Hague claiming intellectual ownership of the telescope. The end result was that no one was granted a patent. Telescopes became the bobble-head dolls of the time. Every street-corner vendor sold them.

What about Galileo? Galileo did NOT invent the telescope, although he may have been the first, or certainly one of the first, to point a telescope at the heavens at a time when most scopes were used to spot enemy ships on the horizon. Galileo's major contribution came in perfecting lens-grinding techniques and telescope design. Galileo tells us in his book *The Starry Messenger* that he heard about the telescope around May of 1609. It's not hard to imagine him obtaining a simple model, dissecting it as a medical school student might a cadaver, and discovering for himself the principles on which it worked.

All telescopes collect light though an opening at the front called the aperture (diagram 5-3). In the very first refracting telescopes, the light passed through a convex objective lens, which caused the light rays to converge to a focus. The distance from the lens to the point where the light rays converge, labeled f.l. in the diagram, is called the focal length of the telescope. A concave lens was placed in front of the focus, and the resulting magnification produced was just the ratio of the focal length of the objective divided by the focal length of this concave eyepiece or secondary lens.

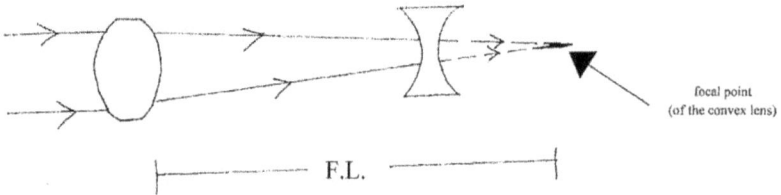

focal point
(of the convex lens)

F.L.

Diagram 5-3

In the "Galilean" design of refracting telescope, a concave "eyepiece" lens is placed in front of the focal point of the convex "objective" lens. The magnification of the scope is just the ratio of the "focal length" of the objective to the focal length of the eyepiece. The main drawback of the Galilean design was its narrow field of view.

Galileo made three telescopes of this design, the last by his reckoning, magnified objects thirty times. With these he discovered the moons of Jupiter, the rings (although he didn't know they were rings) of Saturn, and the phases of Venus. One of the major drawbacks of the Galilean designs was the extremely narrow field of view, or tiny patch of sky visible through the scope.

Galileo's contemporary, Johannes Kepler, the man who set the planets in their proper elliptical orbits (see Chapter 1), came up with a solution for the problem, although he left the construction to others. The Keplerian design replaced the concave eyepiece lens in front of the focus of the objective with another convex lens placed behind the focus of the objective. The result was a much larger field of view, but the resulting image was now upside down (diagram 5-4). This is not of much consequence if your target is the Moon or a nebula, but if you wanted to use the scope for terrestrial viewing it could be somewhat disconcerting.

Diagram 5-4

In the "Keplerian" design the concave eyepiece lens in front of the focus of the objective lens is replace by another convex lens behind the focus of the objective. Kepleriean scopes had a wider field of view, but now the image was upside down.

There were other problems with the refracting telescope. It was the mathematician Rene Descarte who showed that the proper shape for a lens was a curve called a hyperbola. In the seventeenth century it was hard enough to make spherical lenses. Spherical lenses however, suffer from an annoying problem called spherical aberration (diagram 5-5). The light passing through the outer edge is bent more than the light passing through the middle portion of the lens. The different rays come to a focus at different places, making it impossible to achieve a clear focus. Images appear fuzzy.

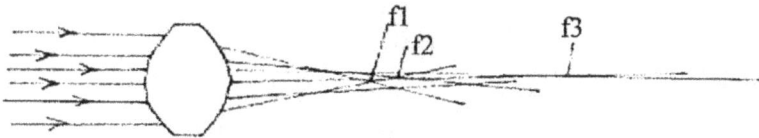

Diagram 5-5

Though a spherically curved lens is easier to grind, light rays hitting the lens at different points are brought to a focus at different points (f1, f2, f3). This flaw is called "spherical aberration." The French mathematician Rene Descarte showed that the proper curve for a lens face was hyperbolic.

The other major problem plaguing early refracting telescopes was inherent in the medium of glass. Lenses are essentially prisms. As such, the different wavelengths of light are bent by different amounts as they pass through the lens, and thus are focused at different places. This is called chromatic aberration and it causes annoying rainbow halos around bright objects, particularly stars.

In the seventeenth century, the only way around these problems was to make objective lenses with a very long focal length relative to the diameter of the lens, or to use astronomer parlance, telescopes with very large focal ratios. This, in turn, meant that the length of the telescope tube had to be very long. Today, when someone talks about a ten-inch or 100-inch telescope, they are talking about the scope's aperture. In the 1600s and 1700s, astronomers talked about 23- or even 36-foot telescopes, but they were talking about the length of the telescope tube. These behemoths had to be supported by crane-like structures and operated by several grown men.

Some astronomers and scope makers decided to forgo the tube altogether, constructing so-called "aerial" telescopes. The objective lens was secured in a short metal tube. The eyepiece was attached to a string, which was pulled tight to achieve the correct focus. Christiaan Huygens constructed one of 210-foot length. Of course, the problem of chromatic aberration could be solved by using a mirror to focus the light instead of a lens. Unfortunately, progress in that arena was slow.

Mirrors had several problems involved in their construction, but the two biggest were shape and smoothness. Descarte also showed that the proper shape for a mirror was a parabola. Spherical mirrors suffer from the same spherical aberration as spherical lenses (diagram 5-6). It was also hard to polish a metal mirror to the required smoothness, and the technology to coat a piece of glass with a reflective surface was years away. Small defects, mountains or valleys, in the mirror surface cause the light to go shooting off in wild directions.

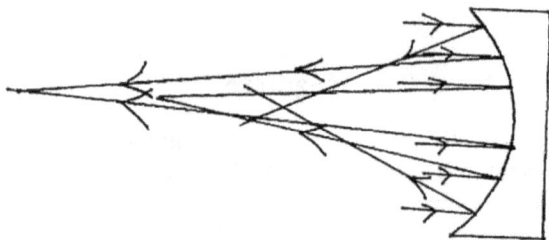

Diagram 5-6

Spherically shaped mirrors suffer from the same problem as spherically shaped lenses. Light rays striking the outside of the mirror are focused nearer the mirror than those striking nearing the mirror's center. The result is a blurry image. Rene Descarte showed the proper curvature for a mirror is parabolic.

Early mirrors were, therefore, all metal. Metal mirrors were hard to cast and had an unfortunate tendency to tarnish. When this happened, the only solution was to remove the mirror and completely re-polish it. Metal also expands and contracts with temperature extremes, which changes the focus of the mirror. In 1616, Jesuit Niccolo Zuchi tried replacing the objective lens of a Galilean telescope with a bronze, concave mirror, but all he saw was a blur.

Isaac Newton experimented with an alloy of tin, copper, and arsenic, the latter added as a whitening agent and to help the metal take a polish. This came to be known as speculum metal. He was also the first to experiment with pitch as a polishing agent instead of cloth or leather. Newton didn't just improve mirror materials. He came up with the telescope design, which bears his name. In the Newtonian design (diagram 5-7), light enters through the aperture and travels to a concave mirror at the back of the telescope tube which reflects the light to a flat mirror suspended from a post at a 45-degree angle near the front of the tube. This flat mirror reflects the light up to the eyepiece lens. As with refractors, the magnification is determined by the ratio of the focal length of the primary mirror to the focal length of the eyepiece. To this day, the Newtonian design is one of the most popular with amateurs and professionals alike. Even with these advances though, the views though Newton's scopes were less than stellar (pardon the pun).

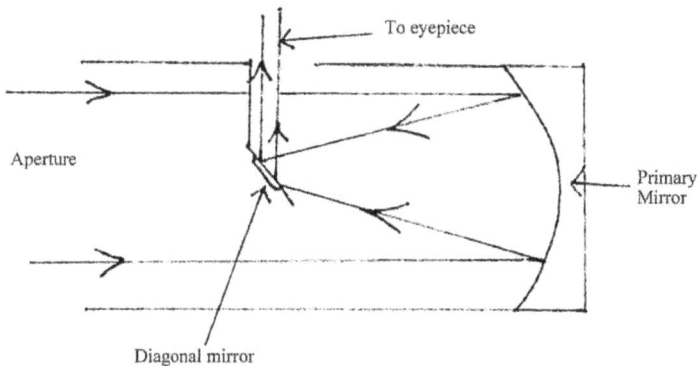

To eyepiece

Aperture

Primary Mirror

Diagonal mirror

Diagram 5-7

In a Newtonian type reflector, light is reflected off a primary mirror to a diagonal mirror, which reflects the light at right angles to the tube out the eyepiece. Newton made his mirrors out of "speculum" metal. Though today the Newtonian design is one of the most popular, the first views were less than stellar.

In 1727, a lawyer and optical tinkerer by the name of Chester Moor Hall devised a solution to the chromatic aberration problem plaguing refractors. The solution involved using two objective lenses instead of one. One lens was concave, the other convex, but the key was that they were made out of two different types of glass (diagram 5-8). Hall chose crown glass and flint glass because they have very different indices of refraction, that is they bend the different wavelengths of light by different amounts. Hall was able to force two of the wavelengths to come to a focus at the same point. Such refractors are called achromatic. Modern refractors can be apochromatic, meaning they bring three or more of the wavelengths to the same focus.

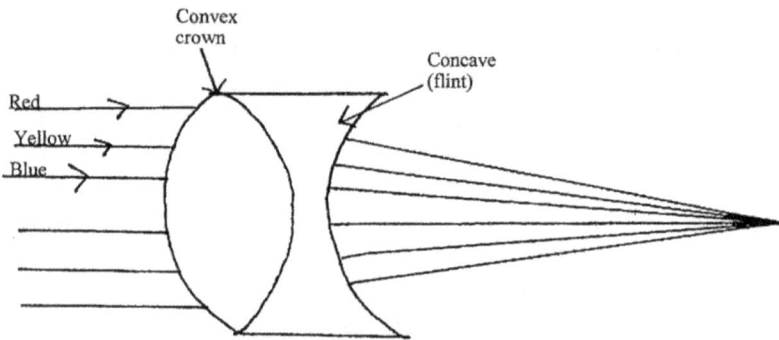

Diagram 5-8

The solution of Chester hall to the problem of chromatic aberration refracting telescopes. A convex and concave lens were paired, but the key is the fact that the lenses are made of different material. The amount by which light is bent by a lens depends on the wavelength, (color) of light, but also the "refractive index" of the lens material. By combining lenses of the right indices of refraction, Hall was able to get rid of chromatic aberration.

Fearful that if he sent the designs for both lenses to the same manufacturer that person would catch on and steal his idea, Hall contracted the work for each lens to a different optician. Unfortunately, both of these professionals were very busy and sub-contracted the work to the same person. Hall continued to believe that an innovation with such obvious benefits to science should be free to anyone who wanted it. He refused to apply for a patent on his design.

The achromatic innovation came a little too late it seems. The era of large-aperture refractors was short-lived. Hall's invention was eclipsed in 1856 by Carl August von Steinheil, who pioneered a technique for chemically coating a concave piece of glass with a thin layer of silver. (Even though aluminum is the material of choice for coating modern mirrors, the process is still called silvering.) Opticians had been casting glass for a long time, and only one side of a glass blank had to be shaped and polished to make a mirror, as opposed to two sides to make a lens. When the surface of the mirror tarnished, instead of re-polishing it, all that had to be done was to re-coat it.

Glass mirrors also weighed much less than metal ones. Glass mirrors were easier to support. Glass did not suffer from extreme changes with temperature like metal. These advantages along with the development by Leon Focault of a relatively easy way (still used today) to test the shape of glass blanks before they were coated made glass mirrors the undisputed choice for large aperture telescopes.

Telescope designs, which had been nothing more than theoretical scribblings on pieces of paper or black board, were now possible. One such design was that proposed by Laurent Cassegrain. In a Cassegrain telescope (diagram 5-9), light enters the aperture and bounces off a parabolic primary mirror at the back of the scope. From there the light is directed to a hyperbolic secondary mirror supported by a truss at the front of the scope. Finally, the light passes through a hole in the primary to the eyepiece. In the 1930s Voldemar Schmidt realized that by adding a curved piece of glass, a corrector plate, to the front of a Cassegrain scope he, could correct the path of the light so that a spherical primary could be used instead of a parabolic.

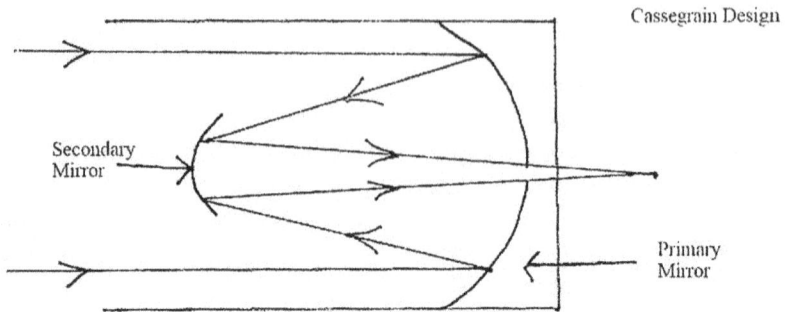

Diagram 5-9

With the ability to grind and coat glass mirror blanks, all sorts of
telescope designs became possible, which were previously impossible.
In the Cassegrain design, light reflects off a parabolic primary mirror at
the rear of the scope, to a secondary mirror in the middle of the
aperture. Later, Voldemar Schmidt discovered that by adding a piece of
glass (corrector or plate) to the front of the scope, a spherical primary
could be used.

    Problems still remained with the manufacture of large mirrors.
Although it may have been easier than casting metal, casting large glass
blanks was still hard, but grinding them and polishing them was a
challenge. At first, large machines were built for the purpose. Then
someone realized that they could let centrifugal force do some of the
work. A mold is put on a turntable. Glass is melted in the mold as it is
spun at a certain rate. The centrifugal force pushes the glass up the side
of the mold into a spherical shape that is retained as the glass is slowly
cooled.

    Next, the process of chemically coating the glass with silver was
replaced by vacuum deposition of aluminum, pioneered in the 1930s. In
this process a mirror blank is placed in an airtight chamber. After the
chamber is evacuated, a tiny amount of aluminum is vaporized and
allowed to settle on the glass surface. Not only is aluminum cheaper than
silver, it is also less prone to tarnish.

    Due to the semi-liquid properties of glass, large mirrors tend to
deform under their own weight. Large mirrors need to be securely
mounted. Some mirrors were lightened by having a honeycomb pattern
of ribs cast into them. Recently, scientists have developed a way to use
many smaller mirror segments to achieve the effect of one large mirror.
Such designs also make adaptive optics possible.

Ground-based telescopes will always be limited in their capabilities by the Earth's protective atmosphere. In adaptive optics, a laser is aimed skyward. As the atmosphere distorts the laser, the individual mirrors can be adjusted to compensate for the distortion. This technique has allowed some ground-based scopes to produce images almost as good as their space-based cousins. That said, the largest telescope on Earth won't do anyone any good if it's just lying on the ground. The proper mount facilitates the pointing of the scope as well as the tracking of objects for astrophotography.

Telescope mounts come in two types: altitude-azimuth and equatorial. Each has its benefits and drawbacks. The simplest of the two is the altitude-azimuth. It gets its name from the fact that it allows the user to move the scope in altitude, up or down, as well as azimuth left or right (diagram 5-10). Altitude and azimuth coordinates are measured in degrees and parts of degrees. From horizon to horizon the sky encompasses 180 degrees. The point in the sky directly above any given point is called the zenith. The altitude of the zenith is 90 degrees (illustration 5-3). Azimuth is measured in degrees counter clockwise (east) from the point where a line drawn from the zenith through the North Star intersects the horizon.

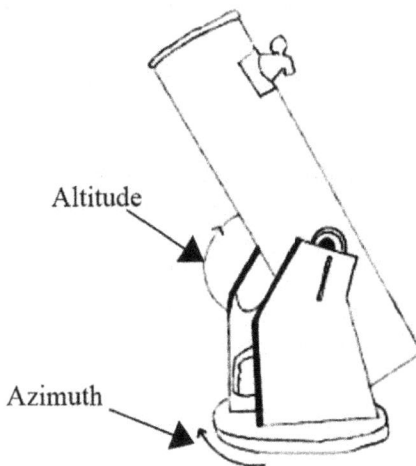

Diagram 5-10

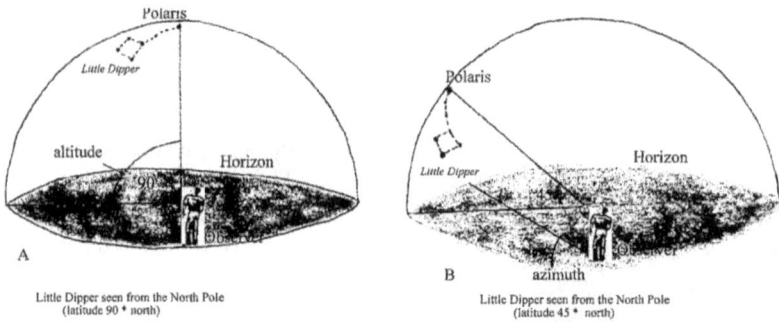

Little Dipper seen from the North Pole
(latitude 90 ° north)

Little Dipper seen from the North Pole
(latitude 45 ° north)

Illustration 5-3

In the altitude-azimuth coordinate system, altitude is measured from
horizon to horizon (0-180) with the zenith (point directly overhead
equal to 90). Azimuth is measured in degrees (0 -360) clockwise from a
point called the North Point.

Equatorial mounts take into account the spin and axis of spin of
the Earth. Equatorial scopes sit on a wedge (illustration 5-4), which can
be raised or lowered to allow for alignment of the polar axis of the
telescope with the celestial north pole. Since the Earth spins with its axis
centered on the celestial north pole, this is equivalent to aligning the
polar axis of the telescope with the rotational axis of the Earth
(illustration 5-5). All that has to be done to keep an object in the field of
view is to turn the telescope in the opposite direction at exactly the same
speed. This allows for long duration time exposure photographs without
blurring or star trails. In modern scopes the turning is accomplished by
high-precision electric motors. Some of the first equatorial mounts were
powered by "gravity drives." Weights were attached to gears and pulleys
and allowed to fall under the influence of gravity, which then turned the
telescope axis at the correct rate.

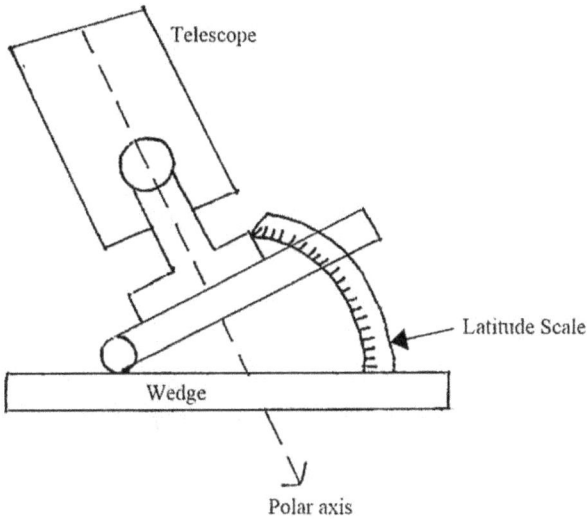

Illustration 5-4

In an "equatorial" mount, the telescope rests on a "wedge." By adjusting the elevation and azimuth of the wedge so that the polar axis of the telescope is parallel to the Earth's rotational axis, the telescope is allowed to track celestial objects. Just turn the telescope at the same rate as the Earth's rotation, but in the opposite direction. This is very important for photography.

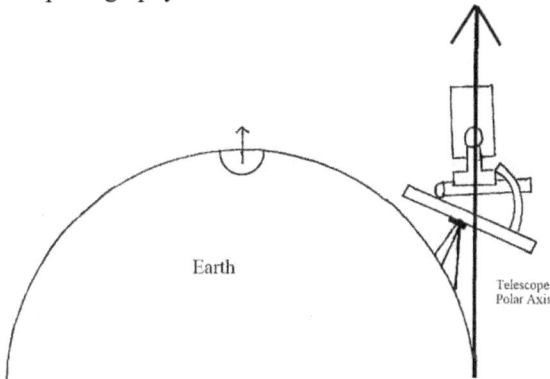

Illustration 5-5

The major benefit of an equatorial mount is that it allows the telescope to be aligned with the Earth's rotational axis. The telescope can then track celestial objects by using a simple motor.

Equatorial coordinates are called right-ascension and declination. These are basically the lines of longitude and latitude used to mark positions on Earth, projected onto the celestial sphere. Like latitude, declination is measured in degrees and parts of degrees north or south of the celestial equator. Like longitude, right-ascension is measured in hours, minutes and seconds, from a line on the celestial sphere equivalent to the Greenwich Meridian on Earth. To find an object with equatorial coordinates 35 degrees, 11h30m simply requires rotating the axes of a properly aligned telescope to those coordinates.

A good telescope, properly mounted also needs the right instrumentation attached to it. For a long time, the only instrumentation available was the human eye. The eye is a wonderful device. They are not very good scientific instruments, however. The unaided eye can only make out objects brighter than 6th magnitude (see Chapter 3). It can discern some color in stars and the brighter nebulae with the aid of a large enough scope, but that's it. Signals go straight from the retina to the brain, do not pass go, with no way to let photons build up for a while before registering. For that we need the camera.

The first astronomical photographs were taken on glass plates coated with an emulsion that produced a negative image when exposed to light. That is, the stars and other bright objects would appear dark on a white background. In combination with devices like the blink comparator (Appendix A), photographic plates made it much easier to detect motion of stars and stellar brightness. Plates taken weeks or even months apart could be compared to look for supernovae (Chapter 3).

Eventually came celluloid instead of glass plates, and then color film. But film had its limitations as well. One of film's major drawbacks was reciprocity failure. In general if you double the exposure time on an object, you get twice the detail in the resulting picture, but past a certain point this is no longer true. The film loses its ability to respond to more light. Scientists were able to develop ways around this failing, but only to a certain point.

The next great advance in image technology came in the computer/electronic revolution. In 1969 two scientists at Philips Research Labs invented the Bucket-Brigade Device, or BBD. One year later two other scientists from Bell Labs improved on the design and the result was the CCD, or Charge Coupled Device, and the rest, as they say, is science and consumer electronics history. Originally designed as a memory device, its light sensitive properties made it a perfect imaging device. One hundred times more sensitive than film, a CCD is an array of metal-oxide-semiconductor capacitors, or charge storage devices, each of which represent a picture element or pixel. The more pixels on a chip the sharper the image.

CCDs are linear detectors. In other words twice the light falling on a pixel produces twice the charge, or signal. Since the image is already in the form of an electronic charge or signal, it can be directly analyzed by a computer. Like film, CCDs can be programmed to accumulate charge for a period of time before discharging, producing time exposure images. CCDs are not trouble free, however.

CCDs are sensitive to thermal noise. The first CCDs had to be cooled with liquid nitrogen. This made them expensive and tedious to use. Recent advances in technology have eliminated the need for such extreme measures, even to the point that amateur astronomers can afford to attach CCD cameras to their telescopes.

The most valuable tool of the astronomer, aside from the telescope itself, is the spectroscope. Most high school level science courses cover Newton's experiments with light and prism. By passing sunlight through several prisms, he was able to show that white light was a combination of the colors of the rainbow, but that the individual colors could not be broken down any farther. In 1802, William Wollaston noticed dark lines in the solar spectrum. He thought these were boundaries between the different colors. Fraunhofer, Bunsen, Kirkhoff, and others discovered that the lines corresponded in position to those observed in the spectrum of calcium, iron, helium and other elements when burned in a flame or placed in a tube and excited by an electrical current in the lab.

In 1862 William Huggins fitted his telescope with a two-prism spectroscope, constructed with the help of his friend William Miller. With this device, Huggins observed similar lines in the spectra of about 50 other stars. The width and brightness of the lines give clues to the star's temperature and the abundance of the various elements. Sometimes, what looks like one line is upon magnification in fact two lines. This phenomena, called Zeeman Splitting, reveals the strength of a star's magnetic field. As discussed in chapters 4 and 6, a shift in these lines from their normal positions indicate motion of the source. All of this information is carried on electromagnetic waves. But there is more to the electromagnetic spectrum than just those wavelengths visible to the eye.

The whole point of science is to remain objective, to keep an open mind regarding ideas until such time as the majority of the evidence points one way or the other. As PC as astronomers try to be, however, sadly most are chauvinists of a different type. Far from revealing a fatally flawed character, this chauvinism is the result of being all too human. Vision is such an important human sense that it is hard to accept that the portion of the electromagnetic spectrum that our eyes are sensitive to is a very small part of the complete spectrum. Each of the other wavelengths let astronomers see a vastly different universe.

At lower energies, or at the longer wavelength end of the spectrum, is the infrared, or what you and I perceive as heat. Infrared helps police see the bad guy in dark urban areas and soldiers to see enemy's body heat in the moonless desert. In the hands of astronomers the infrared is similarly a means to peer into regions and see things that would normally be hidden. Longer wavelengths are not blocked by cosmic dust. Infrared allows astronomers to see stars and planets in the process of forming, or Brown Dwarfs, balls of gas that never quite became stars, that are too cold to radiate in the visible spectrum. Infrared lets astronomers glimpse the early stages of stellar evolution.

At still lower energies we find the radio portion of the spectrum. We take radio for granted as a means of communication between widely separated points on Earth, or even in the case of microwaves as a means of cooking dinner. But radio also provides us with a unique view of the universe. The road to establishing radio astronomy as an accepted science though was not an easy one.

When Karl Jansky, an engineer at Bell Labs, first discovered radio interference at a certain wavelength coming from the Milky Way, the rest of the scientific community took little notice. Of course, that may have had something to do with the fact that Jansky's apparatus looked like something built with a giant erector set.

The first steerable dish antenna was built in 1937 by Grote Reber. Reber used the 31-foot dish to examine sources in and out of the Milky Way. Radio and radio astronomy didn't really come into its own until after World War II. The world's largest single dish is the 1,000-foot diameter Arecibo dish in Puerto Rico. It is literally built into a depression in the landscape. Larger apertures have been made, though, by linking several smaller dishes together.

The Very Large Array telescope is located on the plains of San Agustin near the sleepy town of Socorro, New Mexico. The VLA consists of 27 separate antennas situated on railroad tracks laid out in a Y configuration. Each antenna can be picked up and moved along the Y, producing the effect of a zoom lens on a camera. The VLA also has the distinction of being the most photographed famous telescope in the world, appearing in such movies as 2010, Contact and most recently Terminator: Salvation.

Radio telescopes allow astronomers to explore jets of material from black holes and active galactic nuclei. Radio is also the preferred medium for searching for signals from E.T. (Chapter 6). The broad wavelength of radio does not give details as fine as optical, and the picture is not one that would be recognized in its raw form by the human eye. The signal is analyzed by a computer, which assigns colors to the various signal strengths.

Radio and infrared scopes, like CCD chips, must be cooled to guard against interference from the scopes' own electronics. In addition, infrared scopes must be placed above the Earth's atmosphere to avoid being swamped by heat reflected from the surface of the Earth. Likewise, scopes working in the higher energy/short wavelength regions of the electromagnetic spectrum must also be stationed above Earth's atmosphere. The Earth's atmosphere blocks these wavelengths, otherwise there would be no astronomers to care about them.

Ultraviolet and x-rays give us a unique perspective on the dynamics of our star and others. x-rays are also emitted by some black holes (Chapter 3). If a black hole is near enough to another star, the immense gravity of the hole may pull matter off of the star. As this accumulated matter spirals into the black hole, it loses gravitational energy and also collides with other particles already in the accretion disk, heating it to the point of emitting x-rays. Because of the nature of their quarry, x-ray scopes use a series of open-ended conical mirrors called grazing incidence mirrors. The idea is not so much to reflect the x-rays, but rather to redirect them.

In the 1960s, US military satellites designed to detect Soviet atmospheric nuclear detonations instead discovered short-lived bursts of high-energy gamma-ray photons. Investigation of these phenomena, which could last anywhere from a few milliseconds to several minutes, inspired the launch of the Compton Gamma-Ray Observatory. Current theory predicts that gamma-ray bursts are the result of super-massive stars exploding, or perhaps the collision of super-massive bodies such as co-orbiting neutron stars.

We've come a long way from the telescopes of Lipperhey and Galileo. No doubt there is still a long way to go. Astrophysics is not just a bunch of mathematical scratchings on a chalkboard. A key component, as with all science, will continue to be observation. As the eyes we use to look at the universe evolve, so will our understanding of it.

Sources

Astronomy For Dummies, Stephen P. Maran, Wiley Publishing Inc., Indianapolis, IN,., 2005.

Parallax: The Race To Measure The Cosmos, Alan W. Hirshfeld, W.H. Freeman and Company, New York, NY., 2001.

Stargazer: The Life and Times of The Telescope, Fred Watson, De Capo Press member of Perseus Books Group, Cambridge, MA., 2004.

www.physics.pdx.edu/-d4eb/ccd/index.htm

http://imagine.gsfc.nasa.gov/docs/science/know_11/bursts.html

CHAPTER 6

SEARCHING FOR E.T.

Do you believe in UFO's? This is one of the questions I'm asked most frequently when people discover that I have a degree in physics and an interest in astronomy. (Another would probably be, "What happens to matter when it falls into a black hole?") My answer is an unequivocal, "Yes." I believe in unidentified flying objects. The key word in the phrase being **unidentified**, although if you stop and think about it a bit, the whole phrase is a little oxymoronic. If these things are unidentified how can we be sure they're flying, or if they're truly physical objects? Of course, what people really mean when they ask this question is, "Do you believe that intelligent alien beings have visited Earth in spacecraft?" My answer to that is an equally emphatic, "No." Do I believe there is life elsewhere in the universe? Yes, without a doubt. Intelligent life? Almost certainly. Is there intelligent life elsewhere in the Milky Way? Quite probably. How can I be so sure?

There are a limited number of naturally occurring elements and a finite number of stable combinations of these. While many, in fact most, of these combinations lead to biological dead ends, the sheer vastness of the cosmos argues that life must have occurred in at least one other place sometime during the universe's 13 billion-year history. In fact, scientists have been able to manufacture some of the most basic of biological building blocks in the lab using a few chemicals and electricity to simulate primordial lightning. Basic organic molecules are present in interstellar clouds and on the surface of some comets and asteroids.

For life to form though, it needs a stable platform, a planet with a solid surface. Before E.T. can phone home he/she/it needs a home. So how common are planets? Increasingly, astronomers are coming to the conclusion that the answer is, quite common.

How do astronomers know this? How do astronomers detect planets around other stars? We can't see them (at least not yet). Why? Because planets are small bodies orbiting near their parent stars that shine by reflected light and therefore are very dim. Imagine trying to distinguish the light coming from a firefly hovering next to the beam of a lighthouse from several miles away and you'll have some appreciation of the difficulty of seeing an extra-solar planet.

Astronomers are forced to use indirect methods to discover planets. As the gravity of our Sun pulls on Jupiter, so Jupiter also pulls on the Sun. Indeed, rather than thinking of Jupiter orbiting the Sun it's more accurate to think of both Jupiter and Sun orbiting around a common center. This is true not just of planets and stars, but binary star systems, and black holes and stars (see chapter 3). There are two ways to see the effect of this gravitational ballet.

Theoretically it should be possible to detect the actual wobble, or motion of the star around the common center. This method is called "astrometry." I add the caveat "theoretically" because in practice this method is hampered by several factors. Stars also have a "proper motion" or motion perpendicular to our line of sight that has to be accounted for. The motion due to the pull of a planet is very small. If Jupiter were the only planet orbiting the Sun, the center of their motion would be well inside the body of the Sun. While the Earth's atmosphere is the reason that any of us are here to look for planets in the first place, anyone who has watched shimmering waves coming off hot pavement knows that its turbulent nature makes it very difficult to detect small changes in a star's position.

In order for astrometric techniques to have any hope of finding planets around all but the nearest stars, astronomers must build bigger scopes or find a way to link the ones they have using a technique called interferometry. Throw a rock in a pond and waves spread out in all directions from the site of the impact. The high points are called crests, the valleys of the waves are called troughs (diagram 6-1). Now throw two rocks of roughly the same size into the pond. Where the waves meet they interfere with one another. Where the crest of one wave meets the crest of another (physicists would say the waves are "in phase"), they combine constructively (diagram 6-2). The result is a wave of double amplitude or height. Where the crest of one wave meets the trough of another (180 degrees "out of phase"), they interfere destructively (diagram 6-3). The result is nothing. They cancel each other. The same is true of electromagnetic (light) waves.

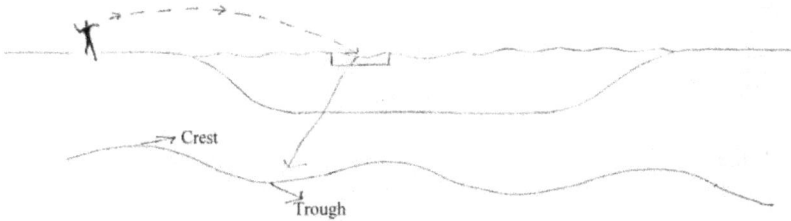

Diagram 6-1

Waves, whether they be water waves, sound waves, or light waves, consist of high points called crests, and low points called troughs.

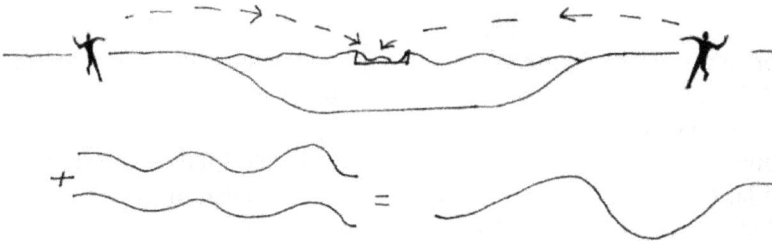

Diagram 6-2

When the trough of one wave occupies the same space as the trough of another, the waves interfere "constructively" and the result is a double amplitude wave (a wave with its crest and trough twice as big as the originals).

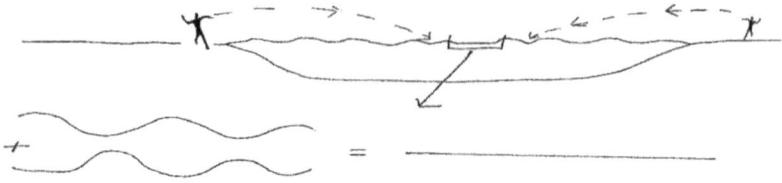

Diagram 6-3

If the trough of one wave meets the crest of another, the result is
"destructive" interference and no wave at all.

By manipulating the light from one or more telescopes so that
they interfere constructively, the resultant image is equivalent to that of a
much larger scope. This technique was pioneered by radio astronomers
because the long wavelengths of radio waves makes them easier to
combine. Recently, however, progress has been made in the optical
spectrum. There are even plans for orbiting interferometers that would be
free of Earth's distorting atmosphere.

The astrometric method works best when the plane of the
exoplanet's orbit is perpendicular to our line of sight. It's also imperative
that the motion of the Earth around the Sun be properly taken into
account. The history of planet searching is littered with the
announcements of the discovery of planets with orbital periods curiously
similar to Earth's only to find on further analysis that the data reduction
failed to take into account Earth's own motion.

Most success in the search for extra-solar planets to date has
been achieved by the spectrographic or "radial velocity" method. In
Chapter 4 we talked about how waves of light experience an analogous
effect to the ordinary Doppler effect experienced by sound waves. When
light is passed through a prism or diffraction grating and is magnified,
the characteristic spectral lines of elements appear. These lines are
shifted toward the blue (shorter wavelength) end of the spectrum when
the source is moving towards us, and toward the red (longer wavelength)
end when the source is moving away. This effect in the light from
galaxies was used by Edwin Hubble to show the universe was expanding.
The same technique can be used to detect the pull of planets on their
parent stars.

The spectrographic technique has its pitfalls as well. The motion
of the Earth must still be taken into account, and because the orbit of
most planetary systems is tilted to our line of sight, the method can only
provide astronomers with a lower limit to a planet's mass, indeed the

method will not work at all if the orbit is perpendicular to our line of sight (as shown in illustration 6-1). That said, it is much easier with present technology to measure small changes in the velocity of a star than it is to detect displacement around a common center.

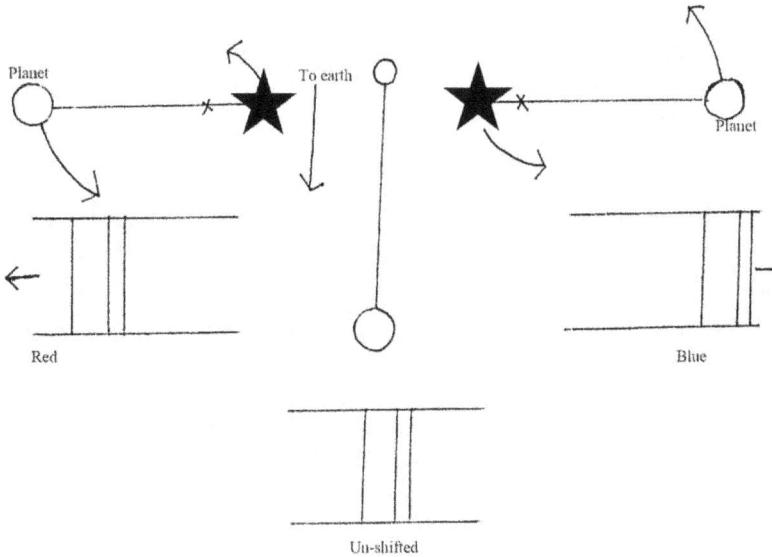

Illustration 6-1

Rather than one body orbiting about another, both revolve around a common center of gravity. As a star and planet do this dance, it is possible to detect shifts in the star's spectra. From the amount of this shift and the mass of the star, it's possible to estimate the mass of the planet.

There are other methods of detecting planets. Every so often (astronomers estimate 1 time out of 100 on average) a planet's orbital plane will be parallel to our line of sight. In that case the planet will periodically transit (see appendices) its star. We can measure this as a periodic dimming of the star's light. The recently launched (2009) Kepler telescope will use this method.

Extreme caution has to be used with this method. Turn a properly filtered telescope towards the Sun on any given day and you're likely to see at least some dark spots on its surface (chapter 3). Just as the

Sun has sunspots, so other stars have starspots. These can mimic the dimming caused by a transiting planet.

None of the methods discussed will allow astronomers to actually see a planet around another star, but here too plans are in the works. What's needed is a way to block out the overpowering light from the star so that dim, reflected light from the planet is allowed to shine through. One way would be to make an actual physical mask, which would be placed in the light path of the telescope, to physically block the light from the star. Another way would be to use a technique known as "null interferometry" or simply, "nulling."

Nulling is the opposite of the process of interferometry discussed previously. Instead of arranging it so that the light from several scopes interferes constructively, the trick is to get the light from the star to interfere destructively, and therefore cancel, leaving only the light from the dim planet. Astronomers at the Very Large Telescope (VLT) in Chile claim to have had some success with nulling.

The results from all of these methods indicate that planets are very common indeed. The number of accepted extra-solar planets is in the hundreds and rising all the time. Even though the majority seem to be oddballs, with masses several times that of Jupiter and orbital periods of weeks to months, most astronomers agree that this has more to do with the limits of our current detection technology than the true nature of the exoplanet population. It's only a matter of time before we start spotting Earth-like planets capable of supporting life.

So if planets are so plentiful, if there are so many potential places for E.T. to call home, then how can I be so sure that E.T. hasn't visited us? Well dear reader, let me count the ways. The main difficulty encountered in any journey between the stars is going to be the distance involved. The Moon is a few hundred thousand miles away. The Sun is 93 million miles away. The nearest star from our Sun is several trillion (yes, trillion with a "t") miles away. To keep from having to say "trillions" all the time, astronomers prefer to talk about light years (see Appendix C). A light year is the distance that light travels in one year, or about 6 trillion miles.

The Milky Way galaxy is about 100,000 light years across. The nearest star to our Sun is three light years away. It would take E.T. three years traveling at the speed of light to reach us. A long trip, but not impossible, and as discussed in Chapter 4, according to special relativity, time for E.T. would virtually come to a stop, but first E.T. has to reach the speed of light. Einstein's equations, specifically $E=mc^2$, says that this is impossible. E.T. can get as close to the speed of light as he/she/it wants, but never 100 percent. As the spaceship speeds up, its mass increases proportionally making it that much harder to gain that last 0.000001 percent. For any massive object to reach the speed of light would require infinite work.

I'll settle for 80 percent, you say. What's a few extra years added to the journey? A very admirable attitude, except that no known propulsion method will get you there. No chemical or solid fuel would work. Not even nuclear propulsion engines that exist only in the imaginations of aerospace engineers will achieve 80 percent the speed of light. But, suppose for the sake of argument that there was a way to achieve a velocity approaching a large fraction of the speed of light. Whoever said getting there was half the fun didn't understand Newtonian dynamics.

When you're in your car, stopped at a stoplight, and the light turns green, what happens? You step on the gas pedal, the car accelerates forward, and you and your passengers are pushed back into your seats. This is a result of Newton's law of inertia, which says (see chapter 1) that objects at rest tend to stay at rest unless acted upon by a force, and objects in motion tend to stay in motion in a straight line unless acted upon by a force. As the car accelerates forward, your body wants to stay at rest. It resists a change in its state of motion and you feel pushed back into your seat.

Physicists measure the effect of acceleration in units of "g's," or multiples of the acceleration on bodies due to Earth's gravity. A body experiencing two g's would weigh twice as much as normal. Shuttle astronauts experience a maximum of 3g's during takeoff. (By contrast, once they achieve orbit astronauts are weightless not because they are in a zero-g environment, but because they are falling at the same rate as their spaceship.) Fighter pilots making tight turns can experience 6 or more g's for brief periods. The human body is not built to withstand the effects of high-g for extended periods of time.

But, let's say that it was, or that some means could be found to negate the effects of the acceleration (in the Star Trek series this is accomplished by devices known as "inertial dampers"). Accelerating at 10 g's starting from rest, it would take almost two years to reach a velocity of 80 percent the speed of light. You can't break the laws of physics, but you might be able to bend them.

The Starship *Enterprise* (can you tell I'm a big fan) used "warp drive" to cover the distances between stars within the time frame of an episode. Remember that according to general relativity matter distorts or warps space-time. By the equivalence of energy and matter given by

$E=mc^2$, so does energy. If you could produce enough energy and use it to selectively warp space-time, scrunch it up in the direction of travel, and stretch it in the opposite direction (see illustration 6-2), then you could travel the distance between the stars without violating Einstein's prohibition against traveling faster than, or equal to, the speed of light.

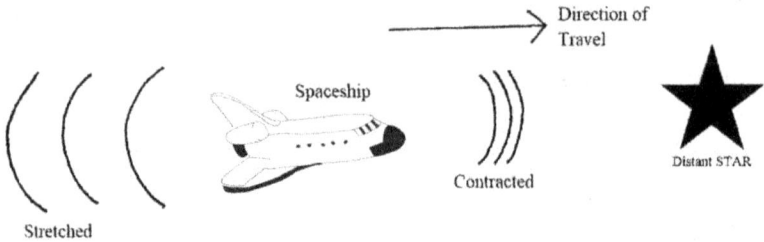

Illustration 6-2

According to Einstein's general relativity energy warps space time. If a way could be found to squeeze or contract space in front of a ship and stretch it out behind the ship, travel to the stars might be possible. This is essentially the "warp drive" envisioned in the Star Trek TV series.

The problem is it would take a tremendous amount of energy. The only substance capable of producing that much energy is antimatter. Antimatter is not a product of the imagination of science fiction writers. It does exist. Antiprotons have the same mass as ordinary protons, but opposite charge (negative in the case of the antiproton). When matter and antimatter are brought together they annihilate each other, and all the mass of both particles is converted to energy. But in order to produce enough energy to warp space-time the way it does, the Starship *Enterprise* would have to be filled from stem to stern with antimatter. There would be no room for transporters, sickbay, photon torpedoes, or the crew that use them.

Another hypothetical way to get around the light speed problem is the ever popular wormhole. Aside from being an undesirable aspect of some apples, wormholes may infest space-time as well. Just like a wormhole in an apple allows the worm to go from point A to point B without slithering over the surface (illustration 6-3), a worm hole in space-time would allow E.T. to go from A to B without covering all the distance in-between (illustration 6-4). There are several problems with this method also.

Illustrations 6-3 and 6-4

Like the wormhole in an apple allows the worm to go from A to B
without crawling on the surface, a worm hole in space would allow a
traveler to get from A to B without covering all the intervening space.

No evidence of a wormhole has ever been found. If they exist,
they probably exist only fleetingly as the tendency would be for the
wormhole to collapse in on itself like a mine tunnel without support
beams. If they do exist they may not get you where you want to go,
rather you may have to go where they take you.

E.T. may have solved the distance problem. Maybe he/she/it has
advanced enough technologically that it can construct wormholes at will.
That still leaves E.T. with the problem of how to breathe, eat, and
excrete. No matter what E.T.'s physiology, no matter what the chemical
basis for E.T.'s construction, there are bound to be some similarities with
Man. E.T. would more than likely have some sort of respiratory system,
whether he/she/it breathed oxygen or some other exotic gas or gases. It's
hard to imagine that E.T. could function without taking in some form of
sustenance, and if E.T. eats, E.T. must excrete. Space on a spaceship
would be at a premium. E.T. would have to find a way of growing
whatever consumables it required in a small space, and would have to
solve the problem of recycling atmosphere, water, and waste.

Then there are the physiological effects of prolonged
weightlessness. We can only guess at the effects on an alien body based
on the documented effects on our own, and these are extremely
detrimental. Space station astronauts exposed to long periods of
weightlessness show increased calcium loss, and marked decrease in
bone density and muscle mass, including the muscles of the heart.
Exercise can slow the process, but nothing has been found that can
substitute for the pull of a gravitational field.

The easiest way to simulate a gravitational field is to set a part of
the spacecraft spinning. When your friend is driving and he takes a turn a
little too fast, you feel as though you're going to be thrown out the

passenger door. This is another example of the law of inertia discussed previously. Your body wants to continue moving in a straight line, but the friction of your posterior in the seat and the closed car door prevent that from happening. As a result you experience a centrifugal force. If a part of the spacecraft is set spinning at the right speed, the result is an acceleration that mimics the pull of gravity. On the other hand, it's just one more level of complexity in the system, one more thing that can go wrong.

Last, but certainly not least, contrary to popular opinion, space is not empty. The region between the stars is filled with gas and dust. So what's a little dust, you say? Nothing, unless the dust collides with you at several thousand miles per second. At those speeds collision with even the smallest particle could be devastating. Development of advanced spacecraft materials or some sort of deflector shield technology would be a prerequisite for any extended journey.

Dust is not the only concern. Space is also filled with harmful radiation: gamma rays, X-rays, cosmic rays. The detrimental effects of this radiation on the DNA of any species goes without saying. E.T. could build hardened areas in his ship to protect against the radiation, but this would significantly increase the mass of the ship.

Still, a good scientist never says never. We thought that the sound barrier was impenetrable. E.T. may have found answers to all these problems, but given the amount of time, energy, natural resources, and what would pass for E.T.'s equivalent of money that such an undertaking would require, I think it's only fair to ask what E.T.'s motives would be in making such a journey. Perhaps E.T.'s planet has suffered some global catastrophe. Maybe E.T. is looking at Earth with an eye to sterilization (destruction of the human population) and colonization. If E.T.'s motives are malevolent, then surely he/she/it must know by now that we pose them no threat. The human race would have been conquered long ago if this were E.T.'s motive. If they want to help us through some precarious period in our development then surely they know the need has never been greater. The way to help would be to contact scientists and engineers who would be in a position to lobby government to implement any technological advance E.T. might have to offer, not by kidnapping loggers on back roads or accountants driving home late at night who think Alpha Centaury is a suburb of Cleveland.

Maybe E.T. just wants to observe. Maybe he/she/it is constrained from making contact by the equivalent of Star Trek's Prime Directive. Surely a race advanced enough to travel the distance between stars with all the inherent difficulties discussed would also have the technology to observe us without being observed.

No one would be happier than I if aliens really were visiting Earth. There are so many questions I would like to ask them. They would have so many advances to share with us. But I hope the preceding discussion has shown you why the odds are stacked heavily against it.

When I get through explaining this, to whomever asked the question, in as much detail as time permits, the next statement is something like, "Okay, Mr. Smartypants. What are UFO's if they aren't spacecraft from other stars or galaxies?" My answer is, "I don't know." This again is the origin of the "U" in UFO. No one knows in all cases what these things are. No one answer explains every sighting. Astronomers have got to get over the fear of admitting that they don't have an answer for everything. The mantra of "swamp gas," or Venus, as an explanation for every UFO has done nothing but fuel the fire of the conspiracy wackos who would have you believe that the government is holding alien bodies at Wright-Patterson AFB, or Area 51.

It may be that bright planets and/or stars and a basic unfamiliarity with the night sky can explain some sightings. As an example, I offer a personal experience. A friend of mine made a frantic phone call to me one evening claiming that he and some fellow workers had seen an object circling the Moon. By the time I got around to looking the object had disappeared, but my friend said he had managed to get some photographs on his digital camera. We belonged to a sci-fi book club and at the next meeting he brought the pictures for me to look at.

I was expecting fantastic pictures with the blurred image of some alien craft racing in circles around the Moon. What I saw instead was two pictures obviously taken several minutes apart that showed a point of light, quite probably Venus or some bright star in two different positions near the Moon. The combined motion of Moon and planet made it appear as though the light was circling the Moon.

I think it's quite probable that a majority of sightings not explained by natural phenomena can be explained by the testing of top secret aircraft by our own or foreign governments. What better way to distract the public and possibly spies from the true nature of the craft than planting stories of crafts from other worlds.

If an alien civilization wants to make contact, there are many easier, and cheaper ways than by building spacecraft to travel the distances between stars. Radio is one way. It's a pretty safe bet that any advanced civilization with an interest in the universe around them will develop radio astronomy at some point in their history. Radio waves, with their long wavelengths, can penetrate the clouds of gas and dust that would, and do, obscure optical scopes and thus can travel long distances. Radio is also a fairly cheap technology, as opposed to lasers.

An artificially generated radio signal should be easy to distinguish from a natural source. An artificial signal should be "narrow band," that is occurring at one particular frequency, not spread over a range of frequencies. If the telescope is pointed away from the source the signal should disappear. It should be detectable by other telescopes around the world. Finally, it should persist for a few hours or days. However, radio does have its own set of problems.

There are infinitely many frequencies on which E.T. might be broadcasting. The first SETI (Search for Extra-Terrestrial Intelligence) program, called Project Ozma, was limited to checking just one frequency at a time. With the continuing improvement of computer technology and telescope instrumentation it's now possible to check millions of frequencies at a time, but we still can't check them all.

It's not enough to agree on the same frequency. In order to communicate with E.T., we must agree on a common language and vocabulary. It's highly unlikely, contrary to most sci-fi literature and movies, that E.T. would step out of the spaceship speaking perfect English. However, one thing that any advanced civilization would have in common is mathematics. What we call Euclidean Geometry they would no doubt call something completely different, but the underlying truths would be the same. In particular, we would most likely share binary arithmetic, the language of computers.

Humans' only message to E.T. was coded in binary and sent to the globular star cluster M13 by astronomers at the Arecibo radio telescope. The message consisted of 1,679 1s and 0s. The key to deciphering the message is the fact that 1,679 is the product of two prime numbers (numbers divisible only by one and themselves): 73 and 23. If E.T. arranges the 1s and 0s as they are received, in 73 rows of 23, he/she/it will find a diagrammatic representation of the telescope that sent the message, the DNA double helix, and a stick human figure among other things. The down side is that M13 is about 5,700 light years away. This means that even if E.T. sends an immediate reply, we won't get it until about the year 13,000 A.D. if there are even any of us still around. Wherever E.T. lives, it's safe to say that instant messaging is not an option.

Of course, the probability of intercepting a message from E.T. is directly related to the number of civilizations that are actively seeking to communicate. Astronomers, even SETI astronomers, like equations just as much as any other scientists. The leader of Project OZMA, the grandfather of modern SETI, Dr. Frank Drake, came up with an equation to estimate how many extra-terrestrials might be trying to communicate. It looks like $N=Rf_pn_ef_lf_if_cL$ where N is the number of broadcasting civilizations. The equation may look daunting with its cryptic nomenclature, but it's actually quite straightforward.

R is the number of solar-type stars around which planets may have formed. Why an emphasis on solar-type stars? Might not planets form around stars more or less massive than the Sun? Certainly, but as explained in Chapter 3, more massive stars burn hot and die quickly leaving a planet with little time to evolve life, much less intelligent life. At the opposite end of the scale, smaller stars might have planets, but the radiant output of such stars might not be enough to allow planetary life to get started.

Planets have been found around oddball stars. One of the first confirmed discoveries of a planetary system was around a pulsar (see Chapter 3). Radio astronomers noticed that what should have been a regular series of pulses coming from the pulsar were not at all regular, coming late sometimes and early others. The only conclusion was that something was pulling the pulsar away from and then towards Earth. Whether the planets formed around the pulsar, or were gravitationally captured after the star went supernova is still a matter of debate. The important thing to note for our current discussion is that the massive amounts of radiation emitted by the pulsar would make life a virtual impossibility on these planets.

F_p is the number of solar-type stars that actually have planets orbiting them. We can only guess at this number, in fact at any of the numbers represented in this equation, but as I've mentioned, we seem to find planets just about everywhere we look. The odds are good, therefore that this number is fairly high.

N_e is the number of planets on which it is possible for life to develop. Several factors enter into play here. The planet must be big enough, that is have enough mass, to retain an atmosphere. The reason we don't find the lighter gases in the Earth's atmosphere, or any gases at all around the Moon, is that what physicists call the "root mean square velocity" (it's not important for you to understand root mean square velocity, but for the interested reader any basic high school or college physics text will have an explanation) of these gases is greater than the escape velocities of these bodies.

Also, a planet must be the right distance from its parent star for liquid water to exist on its surface. This problem is also known in astronomical circles as the "Goldilocks problem." Too close and any liquid water would boil away. Too far and it would freeze. This zone around a star where liquid water would exist on a planetary surface has been named the "habitable zone." For the Sun this zone reaches from about 0.8 Astronomical Unit (the distance from the Sun to Earth, abbreviated AU) out to about 1.6 AU. Venus is too hot, Mars is too cold, but Earth is just right.

More and more scientists are coming to the conclusion that there is also a "galactic habitable zone." The star around which a planet orbits cannot be too close, or too far from the galactic center. Large amounts of radiation near the galactic center might be hazardous to life, yet the star must be close enough for there to be sufficient heavy elements for formation of rocky planets.

F_l is the fraction of these habitable planets on which life has actually formed. Some scientists feel that where the raw materials are present in the proper proportions, life will find a way to establish itself. They point to our own solar system as an example of a case where life may have started on several planets. Numerous spacecraft have now confirmed that water exists in a frozen state in/beneath the soil of Mars. A multitude of surface features indicate that at one time Mars was far more Earth-like, with water coursing over the surface in vast amounts, carving canyons and valleys. Primitive life may well have existed on Mars at this time.

F_i is the fraction of planets on which life has started where that life eventually develops intelligence. Intelligence is in some respects in the eye of the beholder. The late Carl Sagan was a firm believer that dolphins represented a second intelligent species on Earth. In the case of the Drake Equation, we can take intelligence to mean a society capable of inventing radio astronomy for purposes of listening for artificial signals. Where life develops, it's a good bet that given enough time it will gain intelligence and a curiosity about the universe it inhabits.

Finally L is the lifetime of an intelligent civilization. With technology and science comes the responsibility to use them properly, for constructive purposes, not destructive ones. Our civilization has come perilously close to the brink of destruction on at least one occasion (the Cuban Missile Crisis), and perhaps on others that we don't know about. Toxins are seeping into the environment. We are deforesting the planet at an alarming rate. If our own situation bears any resemblance to E.T.'s, they may destroy themselves before they are able to make contact. Or they may decide to get their own house in order before they spend the time and money trying to contact someone else.

At the 1961 SETI conference where Dr. Drake introduced his equation, as the gathered astronomers offered their best guesses as to the values of variables in the equation the terms started to cancel each other out (the product got closer and closer to 1). Eventually the only term left was L. The scientists broke out the champagne and made a toast, "To large L."

For years, SETI as a field of research was considered tantamount to professional suicide. SETI was where PhD's in engineering and physics went to die. SETI researchers were relegated to attaching their equipment parasitically to radio telescopes so they could listen in as "honest" radio astronomers did their work. Although still not considered "cutting edge," SETI is not the graveyard it once was. In the years since Project OZMA, SETI scientists have found a handful of tantalizing signals that satisfied all the criteria given previously for a bonafide signal from E.T. except for the criterion of persistence. In every case, when astronomers turned their telescopes back to the coordinates of the signal, it had disappeared.

At one time in our galaxy's history there may have been several communicating civilizations. They may have destroyed themselves. They may have given up on sending messages and decided to just listen. What a tragedy if there was nothing to hear because everyone gave up sending messages. They may have dismantled their SETI programs because of lack of funding or lack of success. I hope we don't make the same mistake.

The odds of making contact with E.T. may be stacked against us, but the potential benefits are so great that I would hate to see us give up. If we find E.T. then we'll know that a civilization can overcome its petty differences and work for the common good. I have no doubt that we will make contact someday, and on that day the universe will be (I think) a much better place.

Sources

Cosmos, Carl Sagan, Random House, New York, NY., 1980

The Future of Spacetime, Stephen W. Hawking et. al., W.W. Norton & Co., New York, NY., 2002

Is Anyone Out There?, Frank Drake and Dava Sobel, Bantam Doubleday, New York, NY., 1992

New Worlds in The Cosmos: The Discovery of Exoplanets, Michel Mayor and Pierre-Yves Frei, Cambridge University Press, Cambridge UK, 2003

The Physics of Star Trek, Lawrence M Krauss, Harper Perennial, New York, NY., 1995

Website: www.daviddarling.info/encyclopedia/H/habzone.html

APPENDIX A

POOR PLUTO

Generations of children have grown up learning Pluto as the
ninth planet. Then in late August of 2006 the members of the
International Astronomical Union decided at a meeting that poor Pluto
was no longer deserving of the title "planet" and would henceforth be the
primary representative of a new class of object known as the "dwarf
planet" or "Plutoid." The announcement angered and confused many
within the astronomical community, amateur and professional alike, but
whether you agree with the IAU's decision or not there is no debating the
fact that the story of Pluto is a strange one.

In Chapter 1 it was explained that the astronomers Adams and
LeVerier postulated the existence of the planet Neptune to account for
discrepancies in the orbit of Uranus. In 1846, astronomers discovered
one of Neptune's moons. The orbital dynamics of the moon allowed
them to make an estimate of Neptune's mass. The number they came up
with was not enough to account for all the discrepancies in the orbit of
Uranus. This left open the possibility of a body beyond Neptune, a Planet
X, that would account for the rest of the mass needed to disturb Uranus
from its calculated orbit.

Enter into the picture, Percival Lowell. Lowell was one of a
number of amateur astronomers, like William Herschel (see Chapter 1)
who, though they were amateurs, have nevertheless made lasting
contributions to the science. Lowell was born in 1876 to a well-to-do
New England family. In 1894 Lowell used his own money to finance
the building of the observatory outside of Flagstaff Arizona, which
bears his name.

Lowell is remembered in the popular literature for his obsession
with the planet Mars. Captivated by reports from the Italian astronomer
Giovanni Schiaparelli of "canali" or "channels" on the surface of Mars,
Lowell spent countless clear nights at the telescope eyepiece sketching
an intricate series of Martian "canals" which he imagined were carrying

water to a struggling Martian civilization. Try as they might, though, none of Lowell's contemporaries could see what he saw. Lowell gained a reputation, perhaps somewhat undeserved, as a crackpot.

In 1905 Lowell started his own search for Planet X, along with the brothers Vesto and Earl Slipher. The problem was that, unlike Adams and LeVerier, Lowell and the Sliphers had no good idea where to look (aside from somewhere along the ecliptic plane). On top of this, the equipment the astronomers were using for their search was the equivalent (at least measured according to today's technology) of stone tools.

With the development of 35mm film decades away, and digital methods decades beyond that, photos were made using glass plates coated in an emulsion that when exposed to light from a telescope would produce a negative image. That is, stars would appear as dark spots on a white background. Without motorized telescope mounts it was necessary for the human operator to sit at the eyepiece of the telescope in a cold dome (any source of heat would produce turbulence that would destroy the clarity of the image) and guide the exposure by hand so that stars would not produce streaks due to the rotation of the Earth.

Exposures of exactly the same length would be taken of the same area of sky several days apart, under as close to the exact same atmospheric conditions as possible. The two plates produced would then be compared by a carefully trained observer to see whether anything had moved from one exposure to the next.

In 1911, Lowell purchased a piece of equipment that he hoped would help in the search. The device was called a "blink comparator" (illustration A-1). This allowed two plates to be mounted side by side and illuminated from behind. In-between the source of illumination and plates would be a rotating occulting disk that would alternately block the light to one of the plates, then the other, thus "blinking" them. If this was done fast enough the motion of planet (comet, etc.) would become apparent like the motion of a cartoon character when individual frames are run together through a projector at the proper speed.

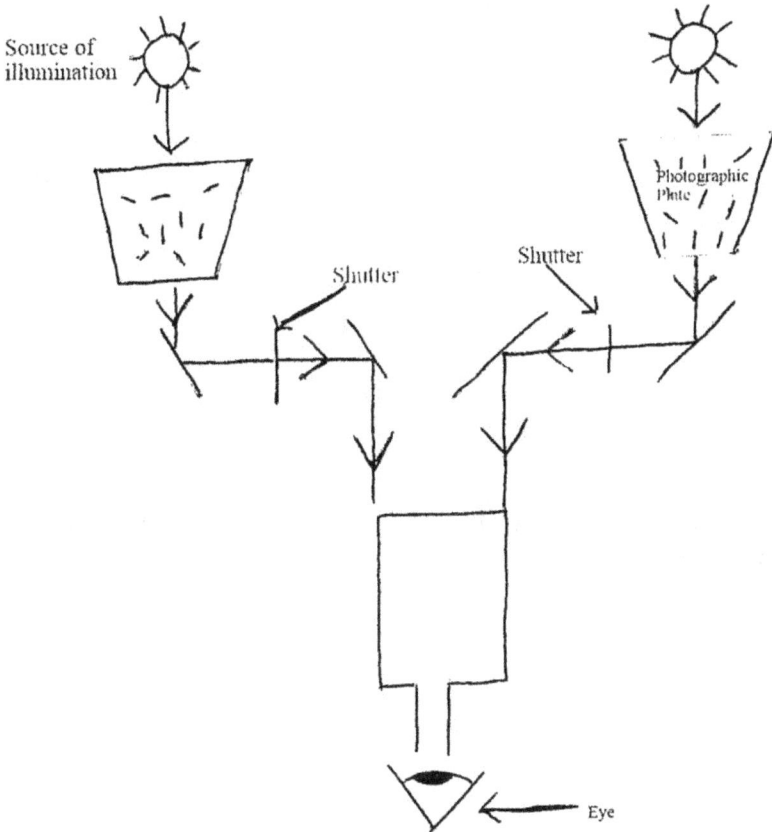

Illustration A-1

Clyde Tombaugh made use of a device called a "blink comparator" or "blink microscope" in his search for Pluto. Two photographic plates of the same area of sky are illuminated and viewed through an eyepiece. By blinking the plates in rapid succession through the use of shutters, motion can be detected from one image to the other.

Unfortunately, Lowell died of a heart attack on November 12, 1916, without having found Planet X. Lowell left a sizable endowment in his will for the continuance of the search. Part of this endowment went towards building a new thirteen-inch aperture telescope for the express purpose of looking for Planet X. The telescope would be of no use, however, without someone to operate it.

For this, Vesto Slipher hired Clyde Tombaugh, son of a Kansas farmer, at a starting salary of $90 per month. Tombaugh arrived at the observatory on January 15, 1929 and on February 15, 1930 he discovered an object moving in the constellation of Gemini. The object was named Pluto after the Roman god of the underworld. It also seemed only appropriate that the first two letters of Pluto happened to be the initials of Percival Lowell. It didn't take long for the oddball nature of the new body to reveal itself.

For one thing Pluto has a very elongated, elliptical orbit. Even though Kepler showed (Chapter 1) that the planets travel in elliptical orbits, rather than circular, for most of the planets the difference between a circle and the planet's true orbit is minimal. Not for Pluto. Astronomers measure the shape of a planet's orbit by its "eccentricity." An eccentricity of zero is equivalent to a circular orbit. The orbits of most of the planets have eccentricities of 0.01-0.02. Pluto's orbit has an eccentricity of 0.25. In fact, the orbit of Pluto cuts inside that of Neptune at one point (see diagram A-1).

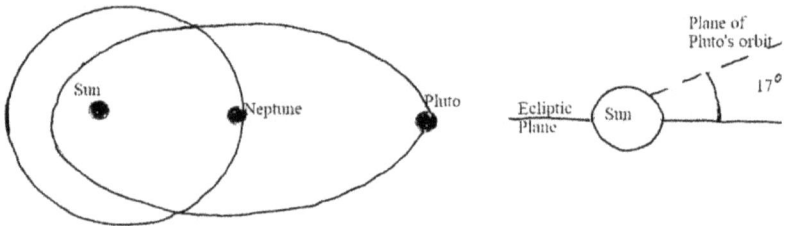

Diagrams A-1 and A-2

Some clues that Pluto is not a planet but rather a member of the Kuiper Belt of objects are its extremely elongated (eccentric) orbit, which even crosses inside the orbit of Neptune; and the extreme tilt of its orbit with respect to the orbital plane of the rest of the planets (the elliptic).

Pluto's orbital period is 248 years. In other words, the Earth goes around the Sun 248 times for every one time Pluto goes around the Sun. At its farthest point from the Sun in its orbit, called aphelion, Pluto is 39.5 times the distance of the Earth from the Sun (93 million miles). Pluto's orbit is also inclined at an angle of 17 degrees to the plane of the orbits of the rest of the planets (see diagram A-2). In short, Pluto's orbit is much more like that of a comet, than a planet.

Pluto also seems out of place with respect to its position from the Sun. The rocky planets (Mercury, Venus, Earth, Mars) are close to the Sun, followed by the asteroid belt, then come the giant gaseous planets. Out beyond the gaseous planets is icy, rocky Pluto. How did a rocky body get out amongst the large, gaseous planets?

Lastly, Pluto is an extremely tiny body. Its mass is only 0.02 of Earth, just 0.5 of Mercury. Hardly enough to account for the extra perturbations of Uranus's orbit. In fact, the most recent estimates of the mass of Neptune provided by the Voyager spacecraft say there is no extra mass needed. There was never any need for Planet X.

Any way you cut it, Pluto is definitely the black sheep of the family. Nevertheless, it remained on the list of planets dutifully learned by many grade school students. Then in a meeting on August 24, 2006, the members of the International Astronomical Union decided to make Pluto the first member of a new class of objects called "dwarf planets." The definition of dwarf planet is, "A celestial body orbiting the Sun, massive enough to be rounded by gravity, but which has not cleared its neighboring region of planetesimals and is not a satellite." Clear as mud! As of this writing there are only five members of the class of dwarf planets: Pluto, Ceres, Eris, Makemake, and Haumeia.

After 76 years why did the IAU feel it was so important to act? Would it have really hurt anything if Pluto were allowed to remain a member of the planets? What were a bunch of scientists who ostensibly got their degrees to do research doing spending time deciding whether it was worthy of the moniker "planet," when they should have been worried about important things like whether Pluto has an atmosphere, or a magnetic field. These are questions you would have to ask the members of the IAU.

Until recently, Pluto had suffered one last indignity. It was the only one of the "classical" planets to remain unvisited by spacecraft from Earth. The Mariner probes visited Mercury and Venus, and three rovers have trod the surface of Mars. Galileo, Cassini, and two Voyager craft have visited the gaseous outer planets. Poor Pluto remained unexplored.

Then in 2006, the same year of the IAU decision stripping Pluto of its planetary status, the New Horizons spacecraft was launched. It won't reach Pluto until 2015. Until then, it will spend most of the journey in "sleep mode." Mission controllers won't wake it up for good until 2014 to check the health of the instrumentation one last time, and send the craft final instructions. Once at Pluto, New Horizons won't linger. It will fly close enough to examine Pluto's atmosphere and surface composition, among other things, before continuing out into the Kuiper Belt (see Chapter 2).

And what of the legacy of Clyde Tombaugh? On board New Horizons is a small container carrying a portion of Tombaugh's ashes. He will be the first human to visit the object he discovered. A fitting tribute to an observational pioneer.

APPENDIX B

ECLIPSES, OCCULTATIONS, AND TRANSITS

As discussed in Chapter 1, the heavens are in constant motion. In trying to satisfy the human need to understand how things work and model what he observes in nature, man built orreries, or mechanical models of the solar system. Using human or other power, these devices simulated the relative motions of the planets and their moons. These were the ancient equivalent of planetarium computer programs, like The Sky, Red Shift, and Starry Night.

As astronomers ran their models forward or backwards in time, the planets each moving with their own speeds in their individual orbits, they found that every so often objects lined up. These alignments have always held a very special fascination for man. Currently much awe and even fear is focused on the alignment of the Earth and Sun with the galactic center in December of 2012. Basically, these alignments come in one of three varieties: eclipses, occultations, and transits.

Eclipses are the events that most observers are familiar with. In an eclipse, one body casts a shadow on another. When the Earth comes between the Sun and Moon, the result is a lunar eclipse. Conversely, when the Moon passes between the Sun and Earth, casting a shadow on the Earth, the result is a solar eclipse. Because the plane of the Moon's orbit is tilted with respect to the plane of the Earth's orbit around the Sun (see diagram B-1), eclipses can come in total and partial varieties.

Diagram B-1

If the plane of the Moon's orbit was the same as the Earth's around the sun, there would be one total solar eclipse and one total Lunar eclipse per month. Since it is not, eclipses come in total and partial varieties, with total solar eclipses the rarest of them all.

Lunar eclipses occur much more frequently than the solar kind. Indeed, if the planes of the Moon and Earth's orbits were the same, total lunar eclipses would occur once a month. Because the Earth is so much bigger than the Moon, and therefore casts a bigger shadow, the total phase of a total lunar eclipse lasts much longer than a total solar eclipse. Total lunar eclipses only happen during the full phase of the Moon.

The shadow from any eclipse is composed of two parts, a darker "umbral" shadow, and a lighter "penumbral" shadow. During a total lunar eclipse, the Moon does not appear completely dark, but rather takes on a reddish hue. This is due to the refraction, or bending, of sunlight by the Earth's atmosphere. The different wavelengths or colors of light are bent by different amounts. The red wavelength is the only wavelength bent enough to hit the surface of the Moon.

Total solar eclipses, though they are much more spectacular, occur much less frequently than lunar eclipses. If you don't travel, you will be very lucky to view one total solar eclipse during your lifetime. As the Moon and Earth travel in their orbits and the Earth spins on its axis, the umbral shadow traces a path on the surface of the Earth (illustration B-1). This path is called the "path of totality." Observers on either side of the path of totality will see a partial eclipse.

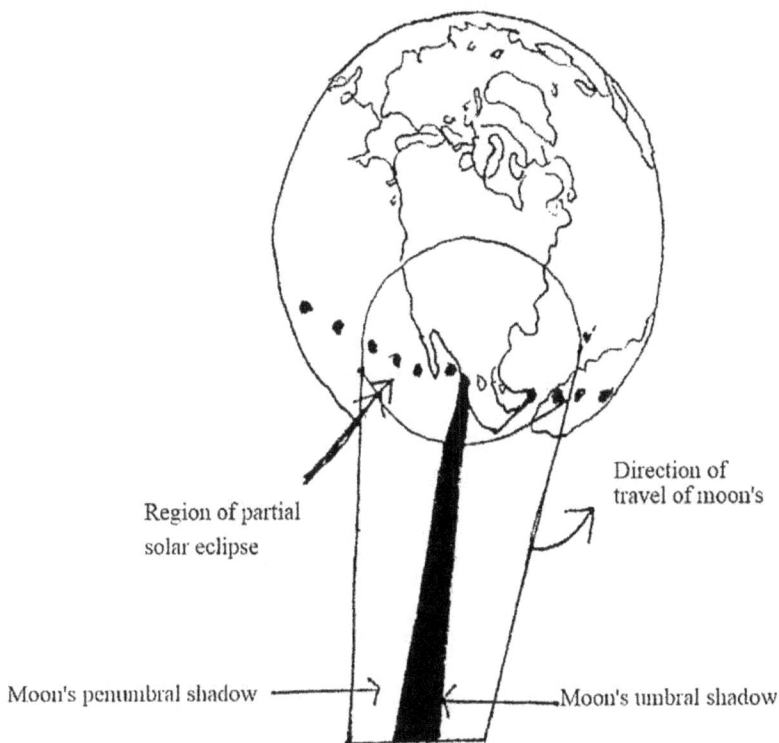

Region of partial
solar eclipse

Direction of
travel of moon's

Moon's penumbral shadow — Moon's umbral shadow

Illustration B-1

How long the total phase of a total solar eclipse lasts for you depends on your location within the path of totality. If the path just grazes your location, totality will be shorter than if you are observing from the middle of the path. It also depends on the size of the shadow, which depends on the Earth-Sun and Earth-Moon distances. Because the Earth and Moon orbit in elliptical paths (see Chapter 1), they are closer to each other at some times than at others. The largest shadow, and therefore the longest totality, would be produced if the eclipse occurred when the Earth was farthest from the Sun (aphelion) and when the Moon was closest to Earth (perigee). An average totality would be about two minutes, but they have been known to last a little more than six minutes.

When the Sun is close, and the Moon is far enough away (illustration B-2), the Moon will not appear large enough to completely cover the Sun. A ring, or annulus, of the solar surface will be visible around the circumference of the Moon. Such eclipses are therefore called annular. The frequency of annular eclipses is increasing as time goes by for a simple reason. The Moon is getting farther away from us.

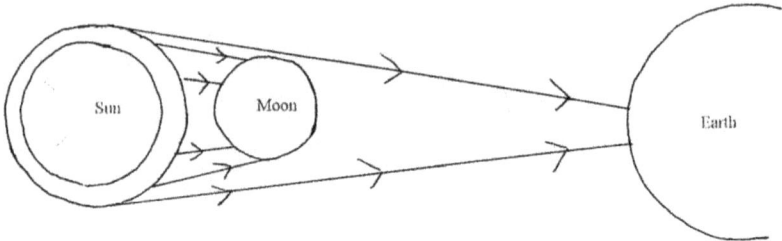

Illustration B-2

When the distance between the Earth, Sun, and Moon is right, the Moon will not appear big enough to black all the sun's rays. The result is an "annular" eclipse, in which a ring of the solar surface is visible around the edge of the Moon. Due to the tidal forces, the moon is moving farther away and annular eclipses are becoming more common.

How is this possible? The Earth is not a perfect sphere. Its spinning motion causes it to bulge at the equator. Therefore the gravity of the Moon pulls unevenly on the Earth, causing the spinning to slow down. The Earth loses some angular momentum. But the laws of physics say the total angular momentum of the Earth-Moon system must be conserved (stay the same). Something must give. The Moon moves farther away by about an inch per year. This has been verified by laser distance measurements. The end result, in a few hundred million years or so, will be no more total solar eclipses, just annular eclipses.

Aside from being pretty to look at, eclipses have and continue to serve their purposes. It's tempting to imagine more than one early explorer with recourse to an almanac, perhaps even Columbus or Magellan, using the occurrence of an eclipse of either variety to get out of a tight spot with the natives (though I don't know if this ever historically happened). Can you imagine the terror as the white man's god makes the Sun disappear or the Moon turn blood red. The total phase of a total solar eclipse is the only time that we can see the outermost layer of the Sun's atmosphere, the corona (see Chapter 3). One particular solar eclipse played a part in confirming a prediction made by one of the greatest physical theories in the history of science.

In Chapter 4, we saw that according to Einstein's general relativity, acceleration is indistinguishable from the force of gravity. Imagine yourself inside a closed container, accelerating at a constant rate (illustration B-3). If you turn on a flashlight, the beam will not hit the wall directly across from you, because in the time it takes the beam to travel the length of the container, the container will have accelerated upward. The beam will appear to have traveled on a curved path (diagram B-2). Gravity then should produce the same effect. It should bend light.

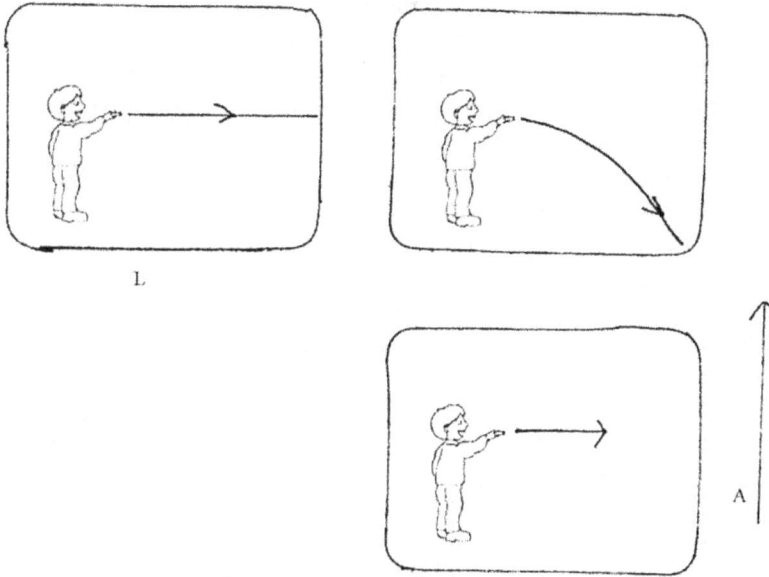

Illustration B-3

If you stand at one end of a room and shine a light at the opposite wall, the beam travels a straight line. If the room accelerates upward at rate A, by the time the beam has gone the length of the room L, the room has accelerated upward. The beam appears to have traveled a curved path. According to general relativity, massive objects should also bend light that passes close to them.

This would only be noticeable near a massive body such as the Sun. But we can only see stars near the Sun during a total solar eclipse. In 1914, an expedition to measure this curvature of light was cancelled due to a little inconvenience called World War I. This was actually fortunate for Einstein, because in the meantime he realized that he had actually underestimated the amount by which starlight would be bent by a factor of two.

A second expedition to west Africa was organized by the British astronomer Arthur Eddington for May of 1919. This eclipse was to take place against the backdrop of the Hyades star cluster. On November 6, 1919, the team announced its findings. Within experimental error, the bending of light was exactly that predicted by Einstein.

An occultation is when one body passes in front of another. Total solar eclipses are also occultations, but the term occultation is generally reserved for events not involving the Earth, Sun or Moon. The Moon can act as the occulting object, however. Lunar occultations are interesting because of the Moon's lack of atmosphere. As the Moon passes in front of a star there is no gradual dimming of the light. The star is simply there one moment and gone the next.

As noted in Appendix D, occultations of Jupiter's moons by Jupiter were used by Ole Roemer to estimate the speed of light. He noticed that the occultations occurred ahead of the predicted time when Jupiter and Earth were on the same side of the Sun, and behind time when the planets were on opposite sides. Dividing the distance of the Earth's orbit by the difference in time of the occultations, Roemer calculated the speed of light.

Stars can occult one another. Certain binary stars are situated such that the plane of their orbit lines up with Earth causing the members of the pair to periodically pass in front of one another as seen from Earth. Such binaries are typically called eclipsing binaries, but a more accurate name would be occulting binaries.

During a transit, a smaller body such as an asteroid or planet passes in front of a much larger body such as a star. From our vantage point in the solar system, the only planets that can transit the Sun are Mercury and Venus. Having witnessed several transits of Mercury, I can tell you they are spectacular events to watch. A tiny, perfectly round black dot, as opposed to the jagged edges of sunspots, gliding across the solar surface is awe inspiring. In fact, Mercury is so close to the Sun that a transit is the best way for many people to see the planet.

The only truly safe way to observe any of these phenomena, at least those that involve the Sun, is by projecting the image of the Sun onto a flat surface. If you decide to look at the Sun through a telescope, be sure that you have a proper solar filter. One of the first telescopes I bought came with a plastic filter that screwed onto the eyepiece. These are very dangerous as they are prone to get hot and crack. A proper filter is one which fits over the front of the telescope and takes out the harmful wavelengths before the light gets magnified.

APPENDIX C

HOW FAR TO THAT STAR?

Many astronomy books written for general audiences try to impress the reader with tables of the distances to celestial objects. That's all well and good. I'm sure quite a few of you reading this book have looked up on a clear night and wondered how far it is to a particular point of light in the night sky. But has it also occurred to you to wonder how astronomers measure these distances? If so, this appendix is for you.

First, let's discuss the units that astronomers use to talk about cosmic distances. Here on Earth we measure distances in meters or kilometers in most countries, or in feet, yards and miles if you live in the United States. The largest of these is good enough for measuring the distance to grandma's house, or from your house to Disneyland. Miles are even fine for talking about the distance to the Moon (about 230,000 miles). For anything much farther than that our usual distance units are just plain inadequate. Astronomers had to come up with a whole new system of units to describe cosmic distances.

For distances within the solar system, astronomers use "astronomical units," abbreviated AU. This unit is the radius of the Earth's orbit, or 93 million miles. Saturn is about nine and one half AU from the Sun. The stars are much farther away than that. They are not millions, not billions, but trillions of miles away. What unit of measure could possibly be adequate for measuring such distances? Light comes to the rescue.

Light, and all other electromagnetic waves travel at the amazing speed of 186,000 miles per second in the vacuum of space. At that speed, light covers a distance of 6 trillion miles in one year. This makes it an ideal measuring stick for large distances. The stars are light-years away. The Milky Way is about 100,000 light-years across. The Andromeda Galaxy is a little over 2 million light-years away. In some circumstances, even light-years aren't big enough. For these occasions there are "parsecs," the derivation of which we will get to shortly, and kiloparsecs (thousands of parsecs) and megaparsecs (millions of parsecs).

Now that we've got that out of the way, how do astronomers actually go about measuring these gargantuan distances? When you want to measure the distance to something here on Earth, you just take out your tape measure and measure it. It's that simple. For larger distances, say the distance to Aunt Bertha's house, you might get in the car. Just note the odometer reading when you start and the reading when you get there, subtract the former and you have your distance. Obviously no such method works for astronomical distances. There was no odometer on the Apollo spacecraft.

Theoretically, the most straightforward way of measuring cosmic distances is the method of trigonometric parallax, or triangulation. As discussed in Chapter 1, parallax is the apparent difference in position of an object as measured against a more distant background when viewed from two widely separated points (diagram C-1). The distance between the points of observation is called the "baseline." Believe it or not, every one of us with normal vision uses this technique every day of our lives. The distance between our eyeballs acts as the baseline. We're not born knowing how to judge distance, but before long the calculation of distance based on parallax becomes second nature to us.

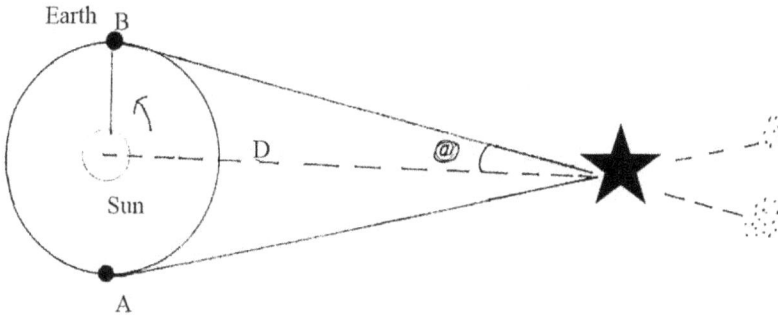

Diagram C-1

The method of trigonometric parallax is conceptually the simplest way to measure the distance to a star. The distance between A and B is called the baseline. By measuring the angle at the distance D can be calculated. In practice, the angle is so small and distortions due to Earth's atmosphere so great, that the parallax is actually a very limited tool.

The problem with measuring celestial parallax is the angle indicated in the diagram is very small. Most of us are familiar with measuring angles in degrees. From horizon to horizon through the zenith (the point directly overhead) is 180 degrees. Smaller measures of angle are the "arc minute" and the "arc second." Each degree consists of 60 minutes of arc and each minute is made up of 60 seconds of arc. An object that exhibits a parallax angle of one arc second when viewed from the extremes of the Earth's orbit is said to be one "parallax second", or one "parsec" away. One parsec is around three light-years.

As if the task of measuring such a small angle isn't difficult enough, there's also the Earth's pesky atmosphere to deal with. Light, any light, even starlight, is refracted or bent upon entering the Earth's atmosphere. Our atmosphere can blur an image, but it can also make a star appear to be somewhere it's not, which is death to a parallax measurement. The other aspect to the atmosphere is, of course, weather. Can you imagine the frustration of making a measurement, waiting six months for the Earth to make half of an orbit only to have clouds obscure your view?

It's kind of a silly example. No self-respecting scientist would rely on only two measurements. Besides the weather, you never know when equipment failures will rear their ugly head. Astronomers take a series of measurements over the course of a year. Taking a series of observations can also give astronomers an idea of where a star resides relative to the plane of the solar system (illustrations C-1, C-2, C-3). If the star traces a circle against the background as Earth makes its orbit, the star is perpendicular to the plane. If it traces a line segment, back and forth, the star lies parallel to the ecliptic. If the shape traced is some intermediate oval, the star is at an angle to the solar system.

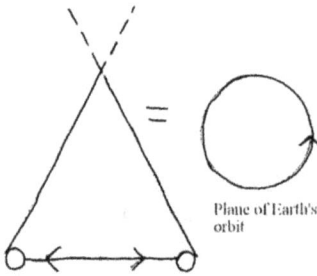

Illustration C-1

Plane of Earth's orbit

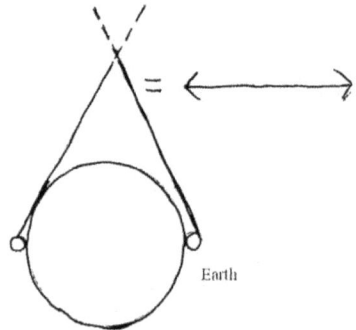

Earth

Illustration C-2

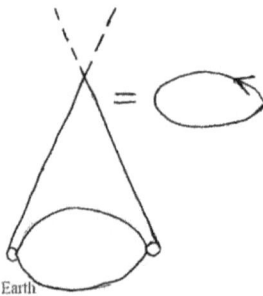

Earth

Illustration C-3

Illustrations C-1, C-2, and C-3

The shape a star traces against the background is an indication of where it lies with respect to the plane of Earth's orbit. If the star is perpendicular to the plane it will trace an almost circular path against the background constellations. If the star is in the plane it will simple appear to move back and forth as Earth orbits the Sun. If the star lies at an intermediate angle, it will trace an oval shape.

With all the difficulties that had to be overcome it's not hard to see why it took time for the technology to progress to the point of being equal to the task of making the first measure of stellar parallax. There is an excellent book on the history of the search for stellar parallax written by Allan Hirschfeld titled <u>Parallax: The Race to Measure The Cosmos</u> that I list in the section on materials for further reading which covers the details of some of the attempts and failures. For the sake of brevity, I'll cut to the chase.

In December 1838, Johan Friedrich Bessel announced that he had used a special type of telescope called a heliometer to measure the annular parallax of the star 61 Cygni. The heliometer got its name because it was designed to measure the angular diameter of the Sun. Whatever its original purpose, it was just what Bessel needed to measure the parallax of 61 Cygni. The heliometer that Bessel used was constructed by premier optician Joseph Fraunhofer (one of the pioneers of the science of spectroscopy) in 1829.

The heliometer was a six-inch aperture refractor with one big difference. The scope's objective lens was split in two. Each half of the lens was secured in its own metal housing. A thumbscrew near the eyepiece housing could be turned, causing the two halves to move laterally with respect to one another. The split objective produced two images in the eyepiece. In effect, the heliometer was two telescopes in one.

Bessel's plan went like this. He would find 61 Cygni and a more distant star in the eyepiece of the telescope. Bessel would turn the thumbscrew until the image of 61 Cygni merged with its more distant companion and measure the offset of the lens halves. Doing this every clear night for several months the parallax of 61 Cygni would thus reveal itself. In the end, Bessel measured a parallax of 0.314 arc second for the star, surprisingly close to the modern value of 0.287 arc second.

In August 1989, the European Space Agency launched the Hipparcos satellite. Unhindered by the limitations of the atmosphere, Hipparcos was able with just an eleven-inch mirror (smaller than some amateur scopes) to measure the parallax of over 100,000 stars. Hipparcos was able to extend the reach of parallactic measurement out to more than 300 light-years. A host of other satellites were to follow: DIVA (Double Interferometer for Visual Astronomy) in 2003, then FAME (Full-sky Astrometric Mapping Explorer) in 2004, SIM (Space Interferometry Mission) in 2005, and GAIA (Global Astrometric Interferometer for Astrophysics) in 2009, each craft improving on the accuracy and reach of the one before. (As of this writing FAME has been cancelled, GAIA has been moved back to a 2012 launch, and I can find no information on the status of either SIM or DIVA.) Even with the best equipment though, there is only so far that parallax will extend. To measure any farther you need another tool.

In a universe where all the stars were carbon copies of one another there would be no need for parallax. The light of a star itself would be our gauge of its distance. Brightness of a light source, like magnetism, gravity, and the electric force, decreases as the square of the distance. (There is a whole school of speculation as to why so many of the forces of nature should behave this way.) As an example, if I know the brightness of a light bulb at a certain distance, and then while my back is turned, or my eyes are closed, you move the bulb farther away, all I have to do is measure the brightness of the bulb at its new distance, and I can calculate the new distance. If the bulb is one-forth as bright as it was before, you have doubled the distance (two times two is four). If the bulb is one-ninth as bright as it was before, you have tripled the distance, and so forth. Unfortunately Mother Nature is not so cooperative.

Not only are the stars not all the same brightness, some stars even vary regularly in brightness. In 1784, John Goodriche discovered a particular type of variable star in the constellation of Cepheus. In 1904, Henrietta Leavitt (Chapter 4) discovered some of these variable stars in the Small and Large Magellanic Clouds, satellite galaxies of the Milky Way. This meant that Leavitt could reasonably assume that all of them, in each cloud at least, were the same distance from Earth. From subsequent observations, Leavitt was able to deduce a period-luminosity relationship for Cepheid variables. Basically, the longer the period, or time it took the star to go from brightest to dimmest back to brightest, the brighter the star. Theoretically with this knowledge, astronomers could measure the distance to a Cepheid and therefore the cluster or galaxy it occupied. Only one problem remained.

Until the scale was calibrated, Cepheids could only provide relative distances. That is, a Cepheid with the same period as another but which appeared one-forth as bright would be twice as far away from us, but that's all that could be said. The key then was to find the distance to at least one Cepheid. But how? The obvious answer would seem to be parallax.

The problem with using parallax to calibrate the Cepheid yardstick was that with the technology available at the time, the diameter of the Earth's orbit was not a long enough baseline. The nearest Cepheid was just too far away. Fortunately the motion of the Earth around the Sun is not the only motion that can be used to create a baseline for parallax measurements. The Sun also moves through the Milky Way, carrying the planets with it. It was our old friend William Herschel (Chapter 1 and Appendix A) who first discovered that the Sun is moving in the direction of the constellation Hercules. All one had to do was find photographic plates taken of a particular Cepheid several years apart, calculate how far the Sun had moved since then (the baseline), and triangulate as usual. Enjar Hertzsprung of H-R Diagram fame (Chapter 3) was the first to use this method to calculate the distance to the Small Magellanic Cloud.

It may not always be possible to spot Cepheids in every galaxy, especially very distant ones. To judge the distance to these, another method must be used. Supernovae (Chapter 3) can briefly outshine entire galaxies. One particular type, called Type 1a, are highly sought after as "standard candles" or indicators of distance. Type 1a supernovae are thought to occur in binary systems (two stars orbiting around a common gravitational center) in which one member is a White Dwarf (chapter 3) and the other is a Red Giant (Chapter 3). The gravity of the White Dwarf pulls matter off the Red Giant. The matter accumulates until the White Dwarf cannot support the added mass and collapses, initiating a supernova explosion. Because all Type 1a supernovae form as the result of a similar process, they all reach a similar peak magnitude (brightness). The trick then is to catch the supernova before it has reached its peak brightness. From the difference between the observed maximum brightness and the theoretical absolute maximum brightness, the distance can be calculated using the inverse square law as in the previous example.

But supernovae are rare events. Even with modern automated searches astronomers may not always be looking in the right place at the right time. The final tool in an astronomer's bag of distance measuring tricks is the expansion of the universe itself. I related in Chapter 4 how Edwin Hubble combined red-shift data accumulated by his partner Milton Humason and Cepheid distance measurements together and came up with a linear relationship between a galaxy's distance from us and the velocity with which it was moving away from us. Written out as an equation it looks like $V=Hd$, where H is a constant term named in honor of Dr. Hubble.

Because of problems calibrating the Cepheid scale early estimates of the Hubble Constant, H, were very large. About 300 km/second/megaparsec. In other words, for every million parsecs a galaxy was distant from us, according to this estimate it would be moving away with a velocity of 300 km/second. In 2003, the first

Hipparcos data for parallax of galactic Cepheids was released, helping to narrow the calculation of H somewhat. Modern estimates place the value of the Hubble Constant around 70 km/second/megaparsec. This allows astronomers to work backwards. If we measure the velocity of a galaxy via its red-shift (its radial velocity), all that is needed is to divide by the Hubble Constant to get its distance from us.

So there you have it. Astronomers measure the universe step by step. Each step on the cosmic distance ladder is constantly refined as new methods become available and are themselves refined by others. Like any ladder, the cosmic distance ladder is only as strong as its weakest rung.

APPENDIX D

WHY IS THE NIGHT SKY BLACK?

Sounds like the type of question a precocious three or four-year-old might ask. To which the typical adult answer would be something like, "What other color would it be?" or "I don't know. It just is, now go bother your sister." All children are born with a burning desire to learn about the world around them. It's how they survive. They do this, to a large extent, by asking seemingly ridiculous questions. Fortunately, at least some children retain this innate sense of wonder and the ability to ask these apparently simple questions despite the mind-numbing, authoritarian nature ("That's the way it is, because I'm your teacher, and I say so.") of our primary school system. A few of these children go on to become scientists.

The most profound questions in science, the ones that end up advancing the cause of science the most, quite often turn out to be the ones that on their surface seem the simplest. Like the best of them, the question, "Why is the night sky black?" is much deeper than it might seem at first glance.

It's a question that is as old as astronomy itself. Many different cultures have come up with their own explanations. Yuri Baryshev and Pekka Teerikorpi, in their 2002 book *Discovery of Cosmic Fractals* tell of a Malayan aboriginal myth of the Moon and Sun goddesses whose children were the stars. They had so many children that the ladies were afraid that mankind wouldn't be able to withstand all the light and heat produced. They agreed then to eat their offspring, an agreement that the Sun goddess kept, but the Moon goddess did not.

Moon goddess simply hid her children and only brought them out at night. This infuriated the Sun goddess who chased the Moon goddess through the sky and continues to do so to this day. Thus, the night sky contains only so many stars and the space in-between is black.

Whatever the ancient Malayans may have called it, modern astronomers know the problem as "Olbers's Paradox" after German astronomer Wilhelm Olbers (see Chapter 2). In its modern formulation, the problem goes something like this: in a universe which is very large, possibly even infinite in extent, where stars are eternal, if the distribution of matter in the form of bright objects (stars and galaxies) is homogeneous (evenly distributed) then in every direction you might look in the night sky your gaze should fall on one of these bright points. The night sky should be as bright as the daytime one. Yes, you say, but what about clouds of obscuring matter? Wouldn't that block some of the light? Yes, but after a while, this dust would itself absorb so much of the radiation from other sources that it would begin to glow. And there is not nearly enough dust to account for the blackness of the night sky.

Olbers lacked a few key pieces to the puzzle. In recent years, it has become apparent that stars and galaxies may not be uniformly distributed in space after all. Astronomers have mapped great masses or clumps of galaxies and also great voids where nothing much of any consequence seems to reside (see Chapter 3). I discussed the lifetime/lifecycle of stars in Chapter 3. Perhaps the most important part of the puzzle Olbers was missing was the finite speed of light.

The nature of light has always been mysterious. Is it a wave or a particle? How is it produced? What medium does it travel in (Chapter 4)? The ancients believed that the human eye emitted rays, which bounced off objects and came back to our eyes to allow us to see. But one of the most enduring mysteries associated with light was, did it have a finite, measurable speed, or was it infinitely fast?

None other than the great Galileo was one of the first to attempt to measure the speed of light. The idea was fairly simple. A pair of his assistants with shuttered lanterns would position themselves on hills several miles apart. One man would open the shutter on his lamp. When his partner saw the light he would open his own shutter (illustration D-1). By measuring the distance between hills and counting pulse beats (remember there were no accurate clocks at the time), the speed of light could be calculated. Indeed there was a measurable delay between signals from opposite hills, but this was entirely due to the reaction times of the individuals holding the lanterns. The speed of light was much too fast to be measured in this fashion.

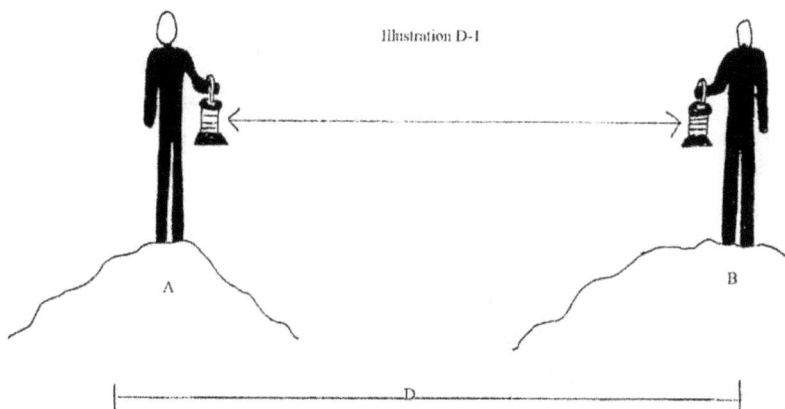

Illustration D-1

Illustration D-1

Galileo's idea for measuring the speed of light was fairly simple. Two assistants with shuttered lanterns positioned themselves on distant hills. The assistant on hill A would open his lantern. When the assistant on hill B saw the light he would open his lantern. By measuring the time (number of pulse beats) from when A opened his shutter to when he saw the light from B and dividing by twice the distance of D, the speed of light could be calculated.

The first reasonable measure of the speed of light is attributed to the Dutch astronomer Ole Roemer who noticed an anomaly in the occultation of one of Jupiter's moons by that body. The occultations would occur ahead of the predicted time when Jupiter was on the same side of the Sun as Earth, and later than predicted when Jupiter was on the opposite side of the Sun. Roemer correctly deduced that the difference must be equal to the amount of time it took light to travel the diameter of the Earth's orbit. This gave the speed of light as 230,000,000 meters/second.

In 1849, the French physicist Armand Fizeau developed a method of measuring the speed of light using a light source, a reflective surface and a cogged wheel (illustration D-2). An observer behind the wheel would either see the light or not depending on whether the reflected light passed through the cogs in the wheel or not. By adjusting the rate of rotation of the wheel, and taking into account the distance from the wheel to the reflector, and the number of cogs in the wheel, Fizeau was able to obtain a value for the speed of light of 310,000,000 meters/second.

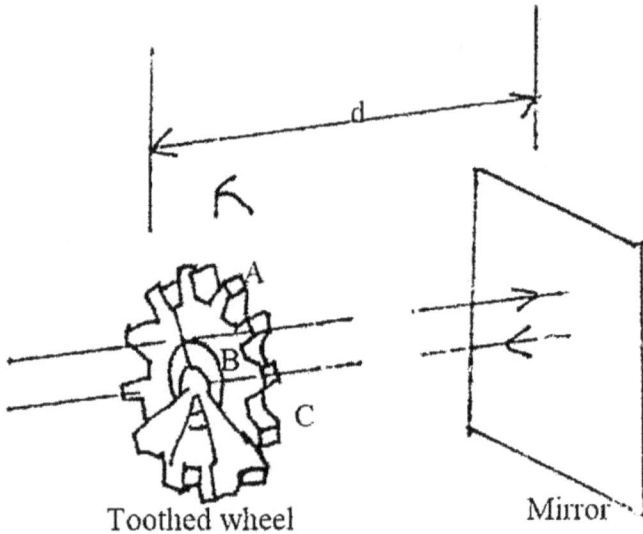

Toothed wheel   Mirror

Illustration D-2

Later in the 1800s, James Clerk Maxwell introduced his equations of electromagnetism. These predicted that all electromagnetic waves should travel at the speed of 299,792,457 meters/second. So far, all experimental measures agree with this number with remarkable precision.

So what does all this mean for our original question of why the night sky is black? Stars are born and die (see Chapter 3), but they weren't all born at the same moment. Once a star begins to shine the finite speed of light means the light from the star takes time to reach us. The distance between stars is so vast that astronomers routinely talk about "light years" or the distance that light travels in one year, or 6 trillion miles (see Appendix C). The light from a star 100 light years from Earth that starts shining as soon as you finish reading this sentence won't reach us for another 100 years. In that amount of time, another star, or more likely several, will have stopped shining because they have used up their nuclear fuel. In the end all these facts (finite speed of light, finite lifetime of stars) guarantee that a sort of balance is maintained and that the night sky is and will remain black.

As with many others, Olbers's Paradox turned out to be not so much of a paradox at all, but rather a result of incomplete knowledge of the universe.

SUGGESTIONS FOR FURTHER READING

As mentioned in the preface, this book is intended to be an introduction to the science of astronomy, a primer if you will. As such, plenty of detail on various subjects, such as the fusion process taking place in different stars, the nature of dark matter (hot vs. cold), and some of the details of cosmology, have been omitted in the interest of brevity. For those of you who have had your curiosity piqued (my whole intention in writing this book) and would like to learn more of the details of certain subjects, or perhaps just get a second opinion on some of the details presented in this book, I present you with the following list of books for further reading.

Some I have used as sources for information presented in this book, some not. Some are fifteen to twenty years old and so are probably no longer in print, but should be available through your local library. All are very well written and imminently readable, at least in my opinion, even by those without a background in math, physics or astronomy. I list them alphabetically by title. Enjoy!

The Case For Mars, Robert Zubrin, Simon & Schuster, New York, NY., 1996.

Coming of Age In the Milky Way, Timothy Ferris, William Morrow & Company Inc., New York, NY., 1988.

Dark Cosmos: In Search of Our Universe's Missing Mass and Energy, Dan Hooper, Smithsonian Books, New York, NY., 2006.

Edwin Hubble: Mariner of The Nebulae, Gale E. Christianson, Institute of Physics Publishing, Bristol, UK., 1995.

Five Equations that Changed The World, Michael Guillen, MJF Books, New York, NY., 1995.

Galileo's Daughter, Dava Sobel, Penguin Books, New York, NY., 2000.

Galileo in Rome, William R. Shea and Mariano Artigas, Oxford University Press Inc., New York, NY., 2003.

Is Anyone Out There?, Frank Drake and Dava Sobel, Dell Publishing, New York, NY., 1992.

Man Looks at The Cosmos: The View From Planet Earth, Vincent Cronin, Quill, New York, NY., 1981.

Miss Leavitt's Stars, George Johnson, W. W. Norton and Company Inc., New York, NY., 2005.

The Neptune File, Tom Sandage, Berkely Books, New York, NY., 2000.

Newton's Gift, David Berlinski, Simon & Schuster, New York, NY., 2000.

Parallax: The Race to Measure The Cosmos, Alan W. Hirshfeld, W. H. Freeman & Company, New York, NY., 2001.

The Physics of Star Trek, Lawrence M. Krauss, Harper Perenial, New York, NY., 1995.

Rocks From Space, O Richard Norton, Mountain Press Publishing, Missoula, MT., 1994.

Stardust, John Gribbin, The Bath Press, Bath, UK., 1995.

Stargazer: The Life and Times of The Telescope, Fred Watson, De Capo Press, Cambridge, MA., 2004.

Thinking Physics: Practical Lessons in Critical Thinking, Lewis Carrol Epstein, Insight Press, San Fransisco, CA.,

Tycho and Kepler, Kitty Furgeson, Walker and Company, New York, NY., 2002.

The Whole Shebang: A State of The Universe Report, Timothy Ferris, Simon & Schuster, New York, NY., 1997.

INDEX

www.ingramcontent.com/pod-product-compliance
Lightning Source LLC
Chambersburg PA
CBHW060607200326
41521CB00007B/682